Serious Managers Guide to AI Guardrails

"Protect Your Company"

"Mitigate Risks"

"Implement AI Responsibly"

Claude Louis-Charles, Phd

Matthew Wilson

Cybersoft Publishing LLC

Fort Washington, MD 20744

Drclaude.net

First Edition March 2026

Table of Contents

1 Introduction

Most organizations didn't decide to become "AI organizations." It crept up on them. A pilot chatbot here, an analytics model there, a vendor tool with a recommendation engine under the hood. Then one day a senior leader asks a simple question you're expected to answer: "Are we sure this thing is safe?" That question is why this book exists.

Serious Manager's Guide to AI Guardrails is for the people who sit in the blast radius of that question—IT leaders, transformation leads, product and operations managers who are accountable for outcomes but are not writing models themselves. You live in the middle: between executives who want AI-powered results and technical teams eager to ship, under regulators who are tightening expectations, and in front of users who assume whatever you deploy is trustworthy. You don't need another abstract AI ethics manifesto or a low-level engineering manual. You need something in between: concrete, manager-ready guardrails that plug into your actual workflows and can survive real deadlines.

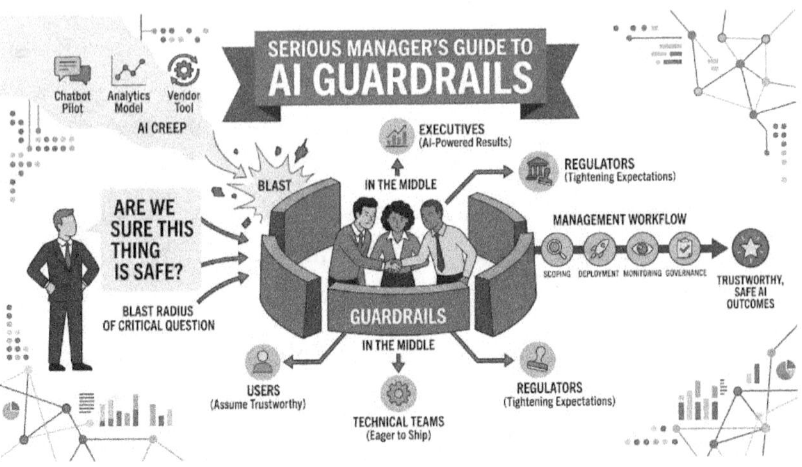

This book starts from a simple premise: AI guardrails are not bureaucratic red tape; they are how you unlock AI at scale without losing control. Guardrails give you operational clarity. They help you decide which AI use cases to pursue, which to park, and which to retire. They help you answer the questions executives will keep asking: Where are we using AI? What risks are we taking? Who is accountable if something goes wrong? How do we know we're not one incident away from headlines or a regulator's letter?

The chapters are organized around the real lifecycle of deploying AI in a modern organization. Early on, you'll see why unmanaged AI quietly accumulates risk in the background—data leakage, bias, brittle models, and one-off exceptions that slowly become the norm. We then move into the backbone of a guardrail program: governance structures, clear decision rights, and workflows that tell teams what "good" looks like without strangling innovation. You'll learn how to translate high-level principles like fairness, transparency, and accountability into concrete steps: what gets checked, by whom, and at what point in the lifecycle.

From there, we go down a level into the mechanics. You'll get practical patterns for technical guardrails that don't require you to be a machine learning engineer to understand. We walk through human-in-the-loop designs that keep humans in command of high-stakes decisions; instead of just "monitoring" automation, they don't have time to challenge. You'll see structured risk triage models that let you treat an internal summarization bot very differently from an automated lending engine—and explain that difference to your board and your auditors.

The middle of the book moves into operational reality: continuous monitoring, drift detection, and incident response. AI doesn't fail like normal software; it degrades quietly as data changes and users adapt. You'll learn how to spot that degradation early, design incident playbooks that teams can execute under pressure, and treat every close call as a chance to harden your guardrails instead of just patching the immediate problem. Alongside that, we explore legal and regulatory readiness—not to turn you into a lawyer, but to give you enough structure to design "compliance once, localize many times" instead of reinventing controls for every new rule.

If you are responsible for technology delivery, you'll recognize the patterns in later chapters: CI/CD pipelines, model registries, testing gates, and monitoring dashboards. What's different here is the framing. We're not bolting governance onto the side of engineering; we're embedding it into the same pipelines that already run your builds and deployments. You'll see how to express guardrails as code, so that fairness checks, data validation, and approval gates run automatically whenever a model or dataset changes. When done right, compliance "just happens" as part of the work your teams already do.

The book ends where most technical discussions never begin: culture and maturity. Guardrails only endure if they become part of how your organization understands professionalism. That means moving from ad hoc reactions after something goes wrong, through emerging and defined processes, to an embedded state where people proactively apply guardrails without being told. You'll get a maturity gauge to locate where you are on that journey and a playbook for moving up a level without pretending you can jump straight to "world class" in a quarter.

A few choices shape how this book is written. First, it is deliberately non-technical in language but technically fluent in content. You won't see code examples for model architectures; you will see the questions you should be asking your teams about data, testing, and deployment. Second, the focus is relentlessly operational. Every chapter is anchored in concrete scenarios, then translated into templates, checklists, and patterns you can take into your own governance boards, program reviews, and sprint planning. Third, responsible AI is treated not just as a compliance obligation, but as a strategic asset. Organizations that can prove their systems are safe, explainable, and well-controlled will be able to deploy more AI more quickly, not less.

This introduction is not a promise that the journey will be easy. Implementing guardrails will surface trade-offs: some projects will slow down, some use cases will be paused or redesigned, and some teams will resist new constraints. But the alternative is "no guardrails, full speed ahead"; the alternative is unmanaged risk that eventually forces you into crisis mode—under scrutiny, out of time, and with fewer options. The point of this book is to help you move first, on your own terms.

As you read, treat this guide less as a linear textbook and more as a toolbox. You might start by using the risk triage model to clean up an existing AI portfolio. Or you might jump straight to the incident response chapter to design a minimal playbook before your first serious outage or bias event. Whatever path you take, keep the core question in mind: if someone asked you tomorrow, "Are our AI systems safe, accountable, and defensible?", would you be able to say "yes"—and show your work? The pages that follow are designed to help you get to that answer.

2 Why Guardrails Matter Now

Opening Scenario: When "Smart" Becomes Fragile

You walk into the Monday stand-up and learn that over the weekend, your new AI assistant confidently generated an email to 30,000 customers. The copy looked polished. The tone matched your brand. Unfortunately, the model also hallucinated a promotional offer your company never approved. Customer support tickets spike. Legal is on the phone. Your CEO asks a simple question: "Who approved this?" Everyone looks at the AI team. The AI team looks at you. In that moment, it becomes painfully clear that "we tested the model on a few prompts" is not a governance strategy. This is where guardrails start—not as theory, but as the visible difference between controlled innovation and reputational exposure.

Guardrails are the structures, processes, and technical controls that keep AI systems aligned with your mission, your policies, and your risk tolerance. They do not make AI slower by default, but they do make it safer, more predictable, and more defensible. For IT managers leading AI modernization, this chapter provides a new lens: guardrails as a strategic enabler, not a compliance tax. You'll see why "just ship the model" is no longer acceptable once AI moves from lab to line-of-business, and why executives will increasingly judge your AI program not only by what it can do, but by how well it is controlled.

2.1 From Experiments to Infrastructure

For years, AI lived in pilots and proofs of concept. Small teams tried new models on curated datasets, often with a researcher or data scientist watching every output. In that environment, risk was localized and mostly theoretical. If something went wrong, the damage was limited to a slide deck or an internal demo. Today, the landscape is different. Generative models sit inside customer-facing chatbots, agentic workflows triage service requests, and decision-support tools shape clinical, financial, and operational choices. AI has quietly crossed the line from experiment to infrastructure.

When technology becomes infrastructure, expectations change. Infrastructure must be reliable, observable, governed, and recoverable. No executive would deploy a payment system without logging, reconciliation, and fraud controls. Yet many organizations are deploying AI with less rigor than they apply to their ERP change controls. This is not because leaders are careless. It is because the mental model of AI has not kept pace with reality. Too many still see AI as "fancy autocomplete" rather than as a probabilistic engine that can influence thousands of micro-decisions per hour. This chapter's goal is to help you reset that mental model.

For IT managers, this shift is especially acute. You are being asked to plug models and agentic workflows into legacy architectures, align them with identity and access controls, and expose them through APIs and integration layers. That is infrastructure work, not experimentation. Guardrails are the bridge that lets you treat AI with the same seriousness as your core systems—without freezing innovation. They define how far an AI system is allowed to go, under what conditions, with which data, and at what level of oversight.

2.2 The Silent Risk Profile of Modern AI

Traditional systems fail loudly. A server crashes. A batch job fails. A transaction is rejected. The alerts are obvious and often technical in nature. Modern AI systems, especially generative ones, fail quietly. They produce plausible but wrong answers. They include sensitive snippets from training data. They phrase content in a way that subtly breaches policy or equity commitments. These failures can remain invisible until someone notices a pattern—or until a regulator or journalist does.

For managers, this "silent risk profile" is AI's most dangerous property. You cannot rely on your usual senses to detect issues. The system appears to be working. Response times are acceptable. User satisfaction may even rise. Meanwhile, the model is slowly drifting, picking up bias from skewed user behavior, or being prompted in clever ways that bypass your naïve filters. Guardrails are the instrumentation and constraints that reveal and contain these silent risks before they become public events.

Silent risks show up in several forms. There is content risk: harmful, inappropriate, or non-compliant outputs being sent to users or employees. There is decision risk: AI nudging or overriding human judgment in ways that systematically disadvantage certain groups or violate policy. There is data risk: models ingesting or exposing information that should never have been used. And there is systemic risk: AI-mediated workflows creating dependencies that are hard to unwind when things go wrong. Guardrails do not remove all these risks, but they make them visible and manageable.

2.3 Why Guardrails Are Not "Bureaucracy."

Many teams resist guardrails because they associate them with bureaucracy. They fear checklists that slow them down, approvals that never come, and committees that say "no" by default. This is a legitimate concern if guardrails are implemented as after-the-fact paperwork. In a modern AI operating model, they should be designed as embedded, automated, and proportional controls that keep teams moving while keeping leadership comfortable with the level of risk.

Guardrails are not about creating new approval empires. They are about clarifying ownership, expectations, and limits. When a manager knows exactly what data a model can access, what types of outputs are blocked, under what conditions escalation is required, and how incidents will be handled, decision-making becomes faster, not slower. The ambiguity disappears. Teams can innovate inside a defined envelope rather than negotiating boundaries for every new experiment.

There is also a morale dimension. Technical teams want to know that they are building something that will stand up to scrutiny. Constantly wondering, "Is this going to get us in trouble later?" is exhausting. A clear guardrail framework signals that leadership has considered risk and is sharing responsibility. That framing changes guardrails from "compliance overhead" to "operational clarity."

2.4 Executive Lens: Trust, Brand & Accountability

Executives do not think in terms of prompt tokens and embeddings. They think in terms of trust, brand, and accountability. When they read about an AI-driven scandal in the news—a model leaking confidential data, a chatbot going off script, an automated decision system discriminating—they ask one question: "Could this happen here?" Guardrails are your answer to that question.

From the executive perspective, three things matter. First, can we explain how our AI systems work at a level that regulators, auditors, and leadership can understand? Second, do we have a clear chain of accountability when something goes wrong? Third, can we show that we made reasonable, responsible efforts to prevent foreseeable harm? Guardrails connect directly to all three. They create evidence: policies documented, risks assessed, tests run, overrides logged, and incidents managed.

This is not only about defense. Robust guardrails create offense. They allow you to take bolder, more visible AI bets because you can demonstrate that you have thought about and mitigated the worst outcomes. In some sectors, this becomes a source of competitive advantage. A bank that can prove its AI credit workflows are controlled and fair will be trusted to automate more decisions. A healthcare provider that can show its AI clinical support tools are monitored and governed can adopt them more broadly. Guardrails build trust, and trust unlocks scale.

From Experiment to Infrastructure

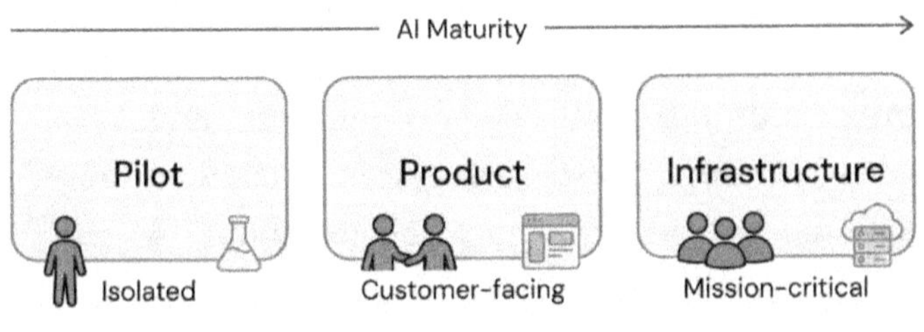

As AI moves right, the need for formal guardrails increases sharply.

2.5 Key Risk Categories Guardrails Address

To design effective guardrails, managers need a simple vocabulary for the types of risk AI introduces. You do not need a PhD in machine learning to do this. You need a clear, shared taxonomy that business, legal, compliance, and technical stakeholders can understand. In this book, we will revisit four broad categories: **content risk, decision risk, data risk, and operational risk**.

Content risk covers everything the AI system generates or surfaces: text, images, recommendations, summaries, and actions suggested to users. Guardrails here might include prompt filters, output classifiers, red-team scenarios, and fallback responses. Decision risk arises when AI influences or makes choices—whether in underwriting, triage, prioritization, or routing. Guardrails here include decision thresholds, human-in-the-loop checkpoints, and bias analysis. Data risk focuses on what the model sees, learns from, and reveals. Guardrails include access controls, data minimization, masking, and robust logging.

Operational risk concerns the system-level behavior: how AI components interact with other systems, how failures cascade, how incidents are detected and handled. Guardrails in this space involve monitoring, alerting, rollback mechanisms, and clear incident playbooks. Having these categories on the table changes conversations. Instead of abstract debates about "AI risk," you can ask concrete questions: "What content risks exist in this use case?" "What decision risks?" "What data and operational risks?" Guardrails become targeted rather than generic.

Four Categories of AI Risk

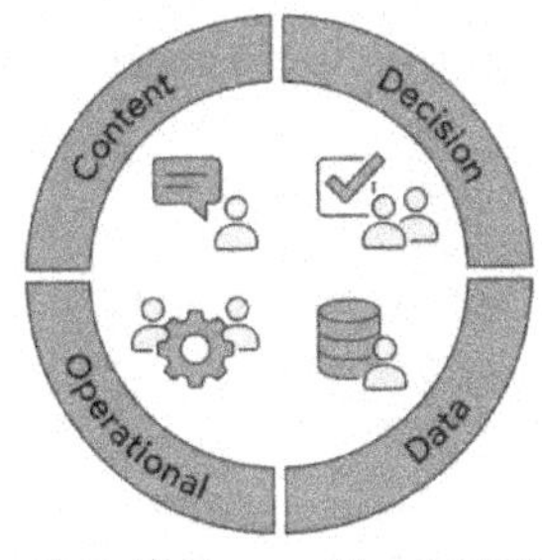

What the AI generates or says.

Choices the AI recommends or makes.

Information the AI uses and exposes.

How AI is deployed, monitored, and scaled.

Use these four categories to structure conversations about AI guardrails.

How Guardrails Enable, Not Block, Innovation

The quickest way to kill an AI program is to say "no" to everything. The second quickest way is to say "yes" to everything and then panic. Guardrails exist to create a third path: controlled experimentation with clear boundaries. When you build guardrails into your AI lifecycle early, you gain the confidence to run more experiments, not fewer, because you know that certain lines cannot be crossed without escalation.

Consider a customer-support chatbot. Without guardrails, your team might restrict its deployment to a tiny slice of FAQs, fearing the reputational risk of free-form answers. With guardrails, you can define specific intents, constrain the model to a vetted knowledge base, implement output filters for sensitive topics, and route unresolved or risky queries to humans. Suddenly, the same organization feels comfortable letting the chatbot handle a much larger portion of inbound traffic. The guardrails unlocked the scope.

This pattern repeats across use cases. In internal knowledge assistants, guardrails around data access and logging enable broader rollout. In decision support tools, guardrails around override and review policies allow more autonomy. In developer copilots, guardrails around code suggestions and security checks reduce fear. The manager's job is to frame guardrails as an innovation accelerator: "These constraints are what make it safe to go faster."

2.6 Where Guardrails Live in the Stack

Managers often ask: "Where do guardrails actually sit?" The answer is: **in multiple layers of the stack**. Some guardrails are policy-level—defining which use cases are allowed, which data sources can be used, and what approval steps exist. Some are application-level—implemented in the orchestration logic around the model: prompt templates, tool access rules, monitoring hooks, and routing logic. Others are infrastructure-level—enforced by identity and access management (IAM), network controls, data catalogs, and logging systems.

Thinking about guardrails only at the model level is a mistake. Models are just one component in a broader socio-technical system. An unsafe prompt design, an overly permissive tool, or a missing audit trail can do as much damage as a mis-tuned model. As an IT manager, you are in a unique position to coordinate across these layers. You can ensure that your IAM policies reflect which applications and services are allowed to call which models, that your logging infrastructure captures the right events for audits, and that your CI/CD pipelines include the necessary checks.

In later chapters, we will dive into each layer. For now, the key idea is that guardrails are not a single product you buy or a script you paste in front of the model. They are an ecosystem of controls, tests, processes, and norms. Your architecture diagrams should show guardrails alongside services, queues, and APIs. When you start to see guardrails as an architectural concern, you will naturally involve

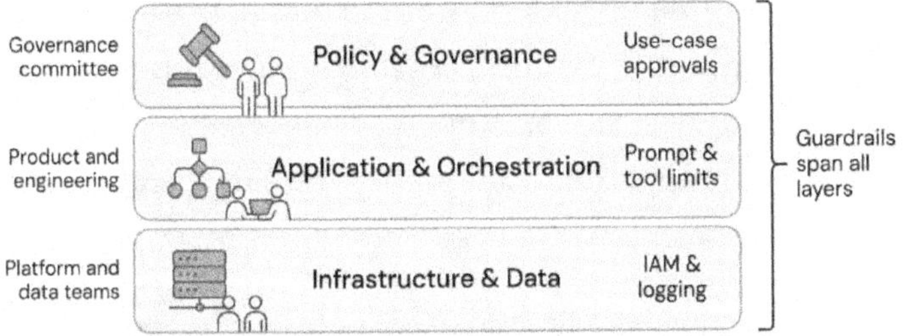

stakeholders early and avoid the "tacked on at the end" problem.

2.7 The Manager's Role in Guardrails

It is tempting to think of guardrails as something legal or security "owns." In practice, if you are leading AI modernization, you will be the one connecting dots. You may not write the policies, but you will translate them into technical controls. You may not design the user flows, but you will ensure that risky steps route to humans. You may not attend every executive meeting, but you will be asked to provide evidence that your systems are under control.

This chapter frames you as the **integrator**. You balance speed and safety. You coordinate across architecture, data, security, compliance, and the business. You build a shared vocabulary for risk. You champion lightweight processes that can scale, rather than ad-hoc heroics. You insist that every high-impact AI initiative answers a simple question: "What are the guardrails, and how do we know they're working?" Over time, that question becomes part of the culture.

You will also manage trade-offs. Not every use case needs the same level of control. A low-impact internal assistant can tolerate more experimentation than an external clinical support tool. Your job is to make those trade-offs explicit, documented, and revisitable. Guardrails should never be "set and forget." They evolve with your portfolio, your risk appetite, and the broader regulatory environment. You are the one who keeps that evolution coherent.

2.8 Why Guardrails Matter Now—Not "Later"

Many organizations treat guardrails as a phase-two problem. "Let's get the model into production," they say, "and then we'll worry about governance." This is understandable in early experimentation, but dangerous once AI touches real users, real decisions, or real data. The cost of retrofitting guardrails rises sharply with scale. The more workflows depend on AI, the harder it becomes to pause and rebuild controls.

There is also a timing gap. Regulations are catching up, but they are not yet fully defined in many jurisdictions. Waiting for perfect clarity is a mistake. The organizations that will be best positioned are those that have already built a responsible-by-design posture. When new rules arrive, they can map existing controls to requirements instead of scrambling from scratch. Guardrails become your hedge against regulatory uncertainty.

Finally, there is a cultural window. As AI programs are still forming, people expect some constraints and learning.

You can establish norms—such as mandatory review of high-impact use cases or default logging of prompts and outputs— without as much resistance as you will face once AI is deeply entrenched. That window will not stay open forever. Guardrails

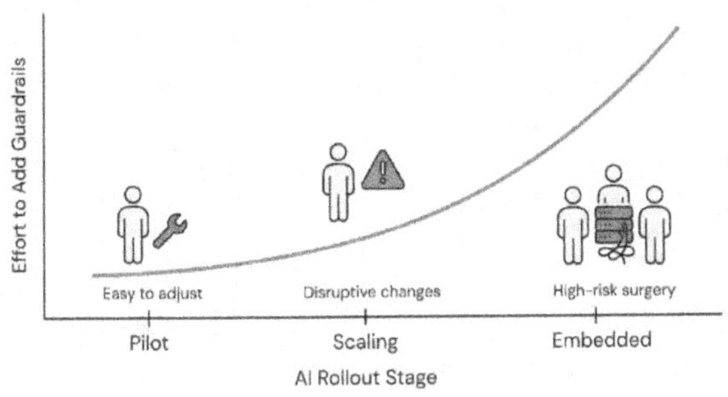

Introduce guardrails early—before AI becomes deeply embedded

introduced early feel like part of how the organization "does AI." Guardrails bolted on late feel like a reaction to failure.

The Manager's Playbook: Starting Guardrails in Your Organization

To close this chapter, we turn to the practical side. As an IT manager, what can you actually do this quarter to move from "AI experiments" to "AI with guardrails"? You do not need a perfect framework. You need a clear starting point. Here is a simple, manager-ready playbook.

First, **inventory your AI touchpoints**. List the models, tools, and agents in use—both official and shadow IT. Identify who owns them, who uses them, and what data they touch. Second, **classify each use case by impact**: internal vs external, advisory vs decision-making vs automated action, low vs high sensitivity. Third, **identify current controls**: Is there any review before deployment? Any logging? Any monitoring? Any clear escalation path? For most organizations, the answer will be "some, but not enough."

Fourth, **define a minimum guardrail set** for each risk tier. For low-risk internal tools, that might mean logging and basic access controls. For high-risk external or decision-making use cases, this will include formal review, human-in-the-loop steps, explicit data restrictions, and periodic testing. Finally, **socialize a simple guardrail policy**—not a 50-page document, but a 2–3-page operational guideline that you can walk through with product owners and technical leads. The goal is to make guardrails concrete, visible, and shared.

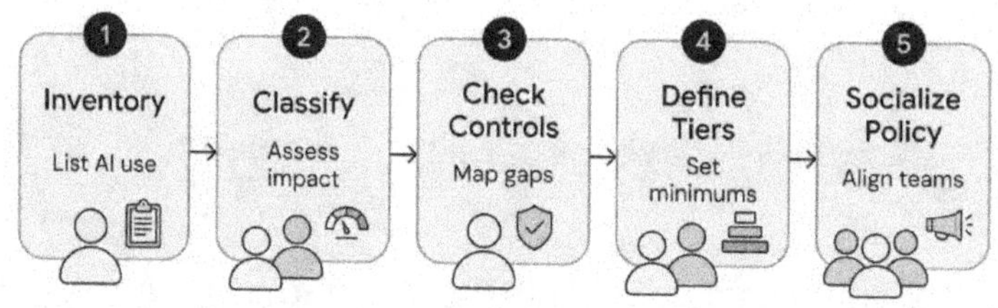

AI Reality Check

Before we wrap, it is worth confronting a few myths that often derail guardrail conversations:

- **Myth 1: "We'll handle guardrails once regulation lands."** Reality: By the time regulation hits, your AI footprint will be much larger, and retrofitting will be painful. Early internal guardrails are your best preparation.
- **Myth 2: "Guardrails will slow us down too much."** Reality: Poorly designed guardrails can, but well-designed

ones speed you up by making boundaries clear and approvals predictable.

- **Myth 3: "The vendor handles all of this for us."** Reality: Vendors can provide features. Only you can decide how those features map to your policies, your data, and your risk appetite.
- **Myth 4: "We don't need guardrails for internal tools."** Reality: Internal misuse or drift can leak data, shape decisions, and train habits that later escape into external products.

Use this reality check with your leadership and peers to reset expectations. Guardrails are not optional extras for AI at scale. They are table stakes.

2.9 Summary: Why Guardrails Matter Now

By now, you have seen why this book opens with guardrails rather than technology. AI has crossed the boundary from experiment to infrastructure. Its risks are often silent but real, affecting content, decisions, data, and operations. Executives care about trust, brand, and accountability. Guardrails, when embedded into your stack and culture, are how you deliver on those concerns without freezing innovation.

As an IT manager leading AI modernization, your role is to integrate, not obstruct. You bring structure to the chaos by inventorying AI touchpoints, classifying risks, and establishing a minimum set of controls that can evolve over time. You help your organization move fast safely. The rest of this book will build on this foundation: translating the idea of guardrails into concrete governance, technical controls, testing strategies, and lifecycle practices. But the core message of Chapter 1 is simple:

> **The right time for guardrails is not after AI is everywhere. It is now.**

3 Foundations of AI Governance

Opening Scenario: The Invisible Owner Problem

Three months ago, your organization launched an AI chatbot to answer employee benefits questions. It was fast, popular, and reduced call center volume. Last week, someone noticed the bot was giving outdated information about parental leave. When HR asked who could fix it, three teams pointed at each other. The data science team said they only trained the model. The application team said they only integrated the API. The product owner said they thought IT was monitoring it. No one had clear authority to update the knowledge base, test the changes, or sign off on redeployment. This is the invisible owner problem, and it is endemic in AI programs that skip governance.

Governance is not about creating bureaucracy. It is about answering simple, critical questions before they become crises: Who decides which AI use cases are approved? Who owns the data pipelines feeding the models? Who is accountable when outputs are wrong or harmful? Who can pause a system, and under what conditions? Without clear answers, even well-intentioned AI initiatives drift into shadow IT, fragmented accountability, and the accumulation of silent risk. This chapter gives you the structures, roles, and templates to build governance that is lightweight, operational, and enforceable.

3.1 What Governance Actually Means in AI Context

Governance is the system of decision rights, accountabilities, and oversight mechanisms that ensures AI initiatives align with organizational goals, policies, and risk tolerance. In traditional IT, governance often focuses on budget control, change management, and vendor oversight. In AI, governance must also address model behavior, data ethics, human oversight, and the fact that systems can drift or degrade without code changes.

For IT managers, this means governance cannot be a separate workstream owned by legal or compliance alone. It must be integrated into how teams propose, design, build, deploy, and monitor AI systems. Governance defines the "rules of the road" for AI: which use cases require executive approval, which data sources are off-limits, what testing is mandatory before production, and how incidents are escalated and resolved. Without these rules, teams either over-ask for permission (creating bottlenecks) or under-ask (creating risk).

Effective AI governance has three layers: **strategic governance** (setting overall direction and risk appetite), **operational governance** (defining processes and standards for day-to-day work), and **technical governance** (embedding controls into infrastructure, tooling, and pipelines). This chapter focuses primarily on the first two layers, while later chapters will address technical implementation. Your role as a manager is to connect all three layers so that high-level principles translate into real constraints in code and workflows.

3.2 Essential Elements of AI Governance Framework

A complete AI governance framework includes six essential elements: decision authority, accountability mapping, policy and standards, review and approval gates, monitoring and reporting, and escalation and incident response. You do not need to build all six at once, but you do need a plan for how each will evolve as your AI footprint grows.

- **Decision authority** defines who can approve new AI use cases, select vendors or models, authorize data access, and set risk thresholds. In small organizations, this might be a single executive. In larger ones, it often involves a tiered structure: a steering committee for high-impact decisions, a working group for operational approvals, and empowered product owners for low-risk experiments.

- **Accountability mapping** clarifies who is responsible for each part of the AI lifecycle: who owns the use case, who owns the model, who owns the data, who owns deployment, and who owns ongoing operations. This is where the invisible-owner problem is solved. A simple RACI matrix (Responsible, Accountable, Consulted, Informed) can prevent weeks of confusion.

- **Policy and standards** translate organizational values into concrete rules: acceptable use policies, data access policies, model approval criteria, testing requirements, and documentation standards. These should be short, clear, and enforceable—not aspirational essays. **Review and approval gates** define checkpoints in the AI lifecycle at which formal sign-off is required. For high-risk use cases, this might include design review, data review, fairness assessment, security review, and go-live approval.

- **Monitoring and reporting** ensure that governance is not just about launch but about continuous oversight. Who reviews model performance metrics? Who gets alerts when outputs drift? How often are dashboards reviewed?

- **Escalation and incident response** define how to handle issues: what constitutes an incident, who investigates, who has authority to pause or roll back a system, and how lessons are captured.

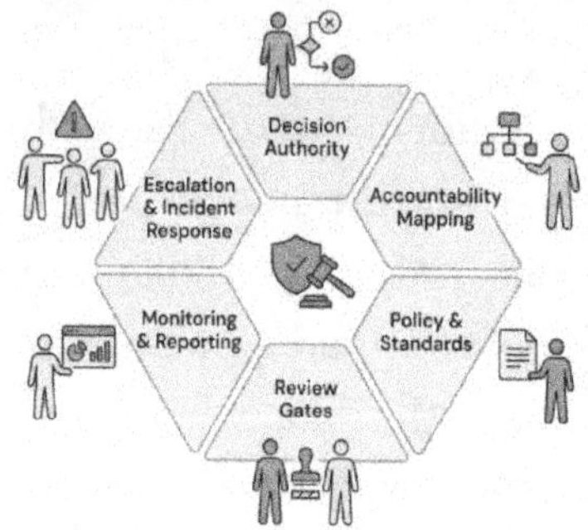

Build all six elements to create durable AI governance.

3.3 Structuring a Multi-Tier AI Governance Board

Most organizations need more than one governance body. A single committee cannot effectively handle both strategic questions ("Should we invest in generative AI for customer support?") and operational questions ("Can we use this new prompt template?"). A multi-tier structure distributes decision-making appropriately while maintaining alignment.

A typical structure includes three tiers: **AI Steering Committee, AI Working Group, and Use-Case Owners**. The **Steering Committee** is executive-level, meets quarterly or semi-annually, and sets overall AI strategy, risk appetite, and investment priorities. Members typically include the CIO, CTO, Chief Data Officer, Chief Compliance Officer, a business unit leader, and sometimes a board representative. This group does not approve individual models; it approves frameworks and high-level direction.

The **AI Working Group** is at an operational level, meets monthly or biweekly, and handles use-case approvals, policy updates, risk reviews, and cross-functional coordination. Members include IT managers, data science leads, product managers, legal counsel, security architects, and representatives from key business units. This group reviews proposals, applies governance criteria, and escalates high-risk or novel use cases to the Steering Committee.

Use-Case Owners are empowered to make day-to-day decisions within approved boundaries. They can iterate on prompts, adjust thresholds, and deploy low-risk changes without committee approval—as long as they stay within the guardrails defined by the Working Group. This tier prevents bottlenecks and keeps teams moving. The key is clarity: everyone knows which decisions they own, which require Working Group approval, and which go to the Steering Committee.

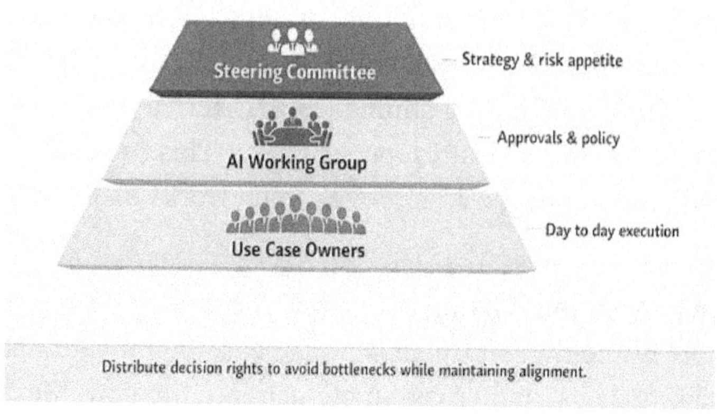

3.4 Roles and Responsibilities: A Practical RACI

Governance fails when accountability is vague. A RACI matrix (Responsible, Accountable, Consulted, Informed) makes roles explicit. For AI governance, key activities include: propose a use case, approve a use case, select a model/vendor, define data requirements, assess risk, design a system, implement controls, test and validate, deploy to production, monitor performance, handle incidents, and update policies.

For each activity, assign roles. **Responsible** means the person or team doing the work. **Accountable** means the single person who owns the outcome and has veto authority. **Consultation** means stakeholders whose input is required before decisions. **Informed** stakeholders must be notified after decisions. A simple example: for "approve high-risk use case," the AI Working Group is Responsible for reviewing and recommending, the Steering Committee is

Accountable for the final decision, legal and security are consulted, and business unit leaders are informed.

A practical RACI for AI should distinguish between **high-risk and low-risk use cases**. High-risk use cases (customer-facing, decision-making, sensitive data) require more consultation and greater accountability. Low-risk use cases (internal tools, advisory only, curated data) can have streamlined approvals. The matrix should also clarify who owns **ongoing operations**, not just the initial launch. Many governance structures focus on "getting to production" and forget about "staying in production safely." Make sure the RACI covers monitoring, updates, and incident response.

Sample AI Governance RACI Matrix

Activity	Use-Case Owner	AI Working Group	Steering Committee	Legal/ Security
Propose use case	R	A	C	I
Approve high-risk use case	R	A	C	
Deploy to production	R	A	C	I
Monitor performance	R	A	C	I
Handle incident	R	A	C	I

Clarify who does what to eliminate the invisible owner problem.

3.5 Frictionless Legal & Compliance

Legal and compliance teams are essential partners in AI governance, but they can also become bottlenecks if not integrated thoughtfully. The key is to involve them **early and structurally**, not as a last-minute approval gate. When legal is consulted only after a system is built, they often say "no" or demand major changes, creating friction and delay. When legal is part of the governance design, they help shape policies and review criteria that teams can apply themselves.

Start by having legal and compliance join the AI Working Group as standing members, not ad-hoc reviewers. This gives them visibility into the pipeline and allows them to raise concerns early. Next, work with them to create **pre-approved templates and patterns**. For example, if legal approves a data access pattern for one use case, document it as a reusable pattern. Future use cases that follow the same pattern can self-certify rather than request a fresh legal review.

Another strategy is **risk-based triage**. Define clear criteria for when legal review is required (e.g., customer-facing, uses PII, makes automated decisions) and when it is optional (e.g., internal assistant, no sensitive data, advisory only). Legal teams appreciate this because it focuses their time on high-impact work. Finally, create a **legal playbook for AI** that answers common questions: Can we use open-source models? What licensing terms matter? How do we handle data residency? What disclaimers are needed? This playbook reduces repetitive questions and builds shared understanding.

3.6 Ownership Models: Who Owns the AI System?

One of the most common governance failures is unclear ownership. AI systems span multiple teams: data engineers, data scientists, application developers, product managers, infrastructure teams, and business stakeholders. Without a clear owner, no one feels accountable for the system's behavior, performance, or risk. Ownership must be assigned and documented.

There are three primary ownership models: **product-led, platform-led, and hybrid**. In a **product-led model**, a product manager or business owner is the accountable party. They define requirements, approve designs, make trade-offs, and own the outcomes. Technical teams (data science, engineering) are Responsible for delivery, but the product owner is Accountable. This model works well for mission-specific AI use cases, such as a customer-facing chatbot or a fraud detection system.

In a **platform-led model**, a central AI platform team owns shared infrastructure, tools, and services (model registry, prompt management, monitoring dashboards), while use-case teams own individual applications built on that platform. The platform team sets standards and provides enablement; use-case teams own compliance with those standards. This model scales well but requires the platform team to have enough authority to enforce guardrails.

A **hybrid model** blends the two: use-case owners are accountable for their applications, but a central AI governance office provides oversight, audits, and escalation. The governance office does not approve every decision but conducts periodic reviews and has the authority to pause systems that violate policy. Choose the model that fits your organization's culture and scale, but choose one explicitly and document it.

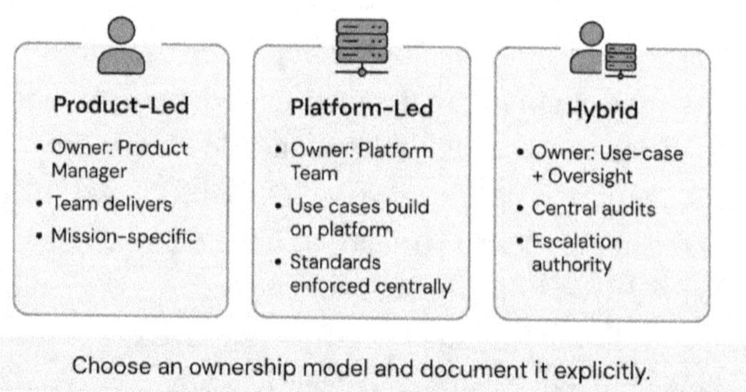

3.7 Escalation Paths and Decision Rights

Governance structures are only as good as their escalation paths. When a novel risk emerges, when two policies conflict, or when a system begins behaving unexpectedly, teams need to know how to escalate quickly and clearly. Escalation paths should be documented, visible, and practiced.

A simple escalation model has three levels: **operational, governance, and executive**. **Operational escalation** handles routine issues: a prompt needs updating, a threshold needs tuning, or a minor bug is detected. These go to the use-case owner or platform team and are resolved within hours or days. **Governance escalation** handles policy questions, risk concerns, or cross-functional conflicts. These go to the AI Working Group and are resolved within days or weeks. **Executive escalation** handles strategic decisions, major incidents, or situations requiring risk acceptance at the highest level. These go to the Steering Committee or C-suite and require urgent attention.

Clear escalation criteria prevent both under-escalation (teams hiding problems) and over-escalation (everything goes to executives). Examples of governance escalation triggers: use case involves sensitive data not previously approved, model performance degrades beyond a threshold, user complaints spike, external inquiries from regulators or the press. Examples of executive escalation triggers: system causes harm to individuals, major data breach, violation of law or regulation, reputational risk to the organization.

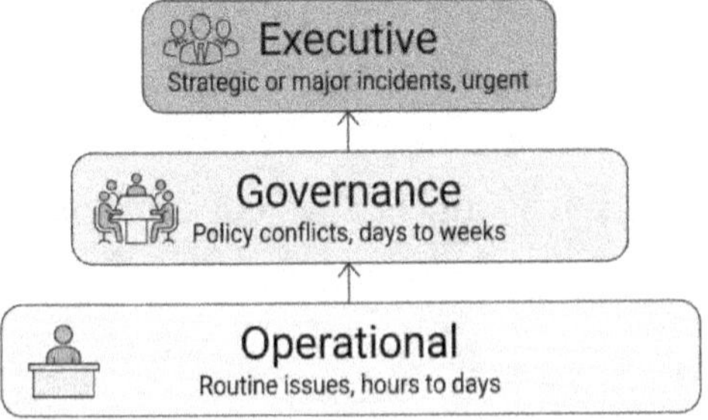

Define escalation triggers so teams know when to raise issues.

3.8 Decision Clarity Without Slowing Innovation

A frequent concern is that clearly defined decision rights may hinder innovation by requiring constant approval. Graduated autonomy addresses this by allowing teams wide latitude within set limits, needing approval only when those limits are exceeded.

Establish a low-risk framework that enables teams to experiment with limited restrictions, including internal tools, advisory-only outputs, pre-approved data sources, no automated actions, and consistent human involvement. Applications within this boundary may be deployed with minimal governance requirements, focusing solely on documentation and ongoing monitoring. For medium-risk activities, implement a streamlined approval process that includes business owner authorization, completion of a security checklist, and a fundamental fairness review. High-risk initiatives must undergo comprehensive governance scrutiny, including approval from the AI Working Group, cross-functional evaluation, and formal documentation of risk acceptance.

This tiered approach balances speed and safety. Teams building low-risk prototypes do not wait for committees. Teams building high-risk systems get appropriate scrutiny. The boundaries should be clear and published. Over time, as teams demonstrate responsible behavior inside the low-risk envelope, you can expand it. The goal is not to eliminate risk but to make risk decisions conscious and proportional.

3.9 The Manager's Governance Playbook

You do not need a perfect governance structure on day one. You need a minimum viable governance (MVG) that you can iterate on. Here is a practical playbook to get started this quarter.

- **Step 1: Draft a one-page governance charter.** Define the purpose of governance (why it exists), the scope (which AI activities it covers), and the key bodies (who decide what). Keep it to one page. Share it with stakeholders for feedback.
- **Step 2: Identify your governance bodies.** If you do not yet have a Steering Committee, start with a Working Group.

Recruit 5–8 people representing IT, data science, product, legal, security, and a business unit. Schedule monthly meetings. If you already have committees, clarify their scope and decision rights.

- **Step 3: Create a simple RACI matrix.** Pick 5–10 key activities (propose use case, approve high-risk use case, deploy to production, monitor performance, handle incident) and assign roles. Use a spreadsheet. Share it with the Working Group and iterate on it.

- **Step 4: Define your risk tiers.** Create three categories: low, medium, and high. Write 3–5 criteria for each (data sensitivity, user impact, automation level, novelty). Document which tier requires which level of approval. This will be your use-case triage tool.

- **Step 5: Document escalation paths.** Write a one-pager: "When something goes wrong, who do you call?" Include names, contact methods, and escalation triggers. Share it with every team touching AI.

- **Step 6: Run a governance dry run.** Take a real or hypothetical AI use case and walk it through your governance process. Identify bottlenecks, gaps, and confusion, and refine your process based on what you learn.

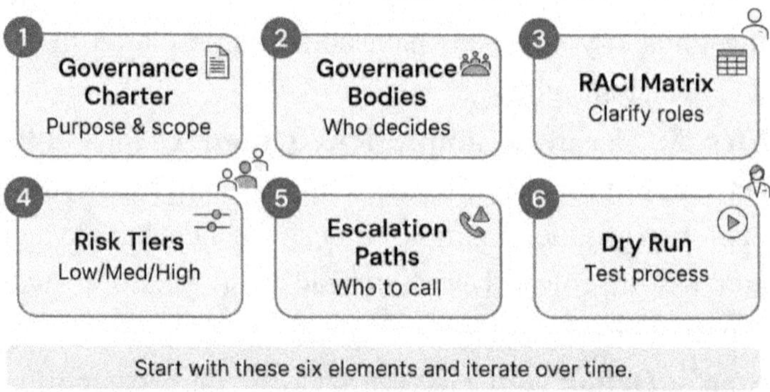

3.10 Avoiding Common Governance Pitfalls

Even well-intentioned governance efforts can fail. Here are five common pitfalls and how to avoid them.

- **Pitfall 1: Governance as theater.** Committees meet, policies are written, but nothing actually changes. Teams bypass governance because they have no teeth. **Avoidance:** Give governance bodies real decision authority. They must be able to say "no" and have it stick. If governance is advisory only, rename it to avoid confusion.

- **Pitfall 2: Over-indexing on policy, under-investing in process.** Beautiful policy documents sit on a SharePoint site, while teams have no idea how actually to get approval. **Avoidance:** Create simple, visual process maps. Make approval workflows accessible. Provide templates and checklists teams can use.

- **Pitfall 3: One-size-fits-all governance.** Treating all AI use cases the same creates either excessive bureaucracy for low-risk work or inadequate oversight for high-risk work.

Avoidance: Use risk tiers and graduated autonomy. Match governance intensity to risk level.

- **Pitfall 4: Governance without technical understanding.** Committees make decisions without understanding how AI systems actually work, leading to impractical or unenforceable rules. **Avoidance:** Include technical experts in governance bodies. Provide non-technical members with AI literacy training. Use plain language.

- **Pitfall 5: Static governance in a dynamic environment.** Governance structures set up once and never revisited become obsolete as AI capabilities, risks, and regulations evolve. **Avoidance:** Schedule quarterly governance reviews. Treat governance as a living system. Iterate based on lessons learned.

3.11   Templates and Tools for Governance

To make governance actionable, you need practical templates. Here are five templates to create or adapt for your organization:

- **Template 1: AI Use-Case Proposal Form.** A simple form that any team proposing an AI use case must complete. Include fields for: use case description, business value, data sources, model type, user impact, risk assessment, proposed guardrails, and owner. This form feeds into your approval process.

- **Template 2: Risk Triage Criteria.** A one-page decision tree or rubric that helps teams and reviewers classify use cases as low, medium, or high risk. Include yes/no questions like: Does it involve PII? Is it customer-facing? Does it make automated decisions? Does it use novel or untested models?

- **Template 3: Governance Review Checklist.** A checklist for the AI Working Group to use when reviewing proposals. Include items like: business case clear, data access authorized, security assessment complete, fairness review conducted, monitoring plan defined, incident response plan documented, and owner assigned.

- **Template 4: Escalation Contact Sheet.** A one-pager listing who to contact for operational, governance, and executive escalations. Include names, roles, email, phone, and Slack/Teams channels. Update it quarterly and distribute widely.

- **Template 5: Governance Meeting Agenda Template.** A standard agenda for AI Working Group meetings. Include standing items: review new proposals, review active issues, discuss policy updates, review incident reports, and assign action items. This creates consistency and ensures nothing falls through the cracks.

Five Governance Templates

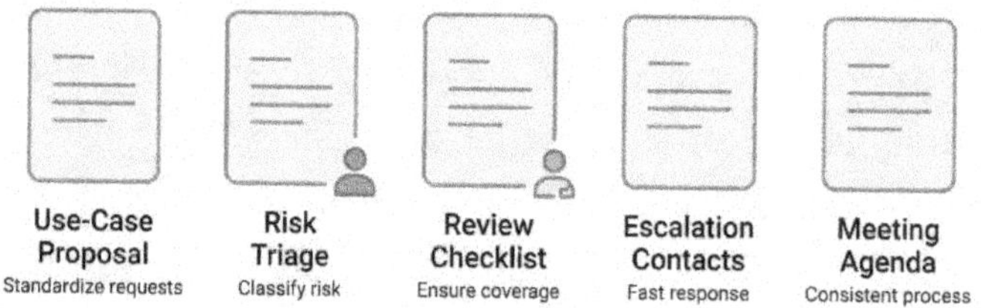

Use templates to make governance operational, not theoretical.

3.12 Summary: Governance as Basis of AI Guardrails

Governance is not a nice-to-have. It is the backbone of any guardrail-based AI program. Without clear decision rights, accountability, and processes, even the best technical controls will fail because no one knows who is responsible for enforcing them. This chapter has given you the building blocks: the essential elements of governance, a multi-tier structure to distribute decisions, a RACI model to clarify roles, integration strategies for legal and compliance, ownership models to eliminate invisible owners, and escalation paths to handle the unexpected.

Your next step is to build your minimum viable governance. Draft your charter, recruit your Working Group, create your RACI, define your risk tiers, document your escalation paths, and run a dry run. Do not wait for perfection. Governance is a living system that improves through practice. The goal is not to eliminate all risk or slow down innovation. The goal is to make AI decisions visible, accountable, and aligned with your mission. When governance works, teams move faster because they know the rules and trust the process. That is how guardrails enable scale.

4 Policy to Practice: Operationalizing AI Principles

Your organization spent six months developing a beautiful set of AI principles. They were word-smithed by legal, approved by the executive team, and published on the corporate website. The principles were clear: AI systems must be fair, transparent, explainable, privacy-preserving, and aligned with organizational values. Three months later, your data science team is about to deploy a customer recommendation engine. You ask, "How does this system demonstrate fairness?" They stare at you. No one has a process. No one has a checklist. No one knows what "fair" means in operational terms. The principles exist, but they have no teeth. This is the implementation gap, and it is one of the most common failures in responsible AI programs.

Having principles is necessary but not sufficient. Principles are aspirational statements about values and intent. They answer the question, "What do we believe?" But they do not answer the question, "What do we do on Monday morning?" This chapter shows you how to bridge that gap. You will learn how to translate high-level principles into repeatable procedures, embed fairness and transparency checks into release gates, tie AI behavior to your code of conduct, and establish regular compliance audits that actually drive behavior change. The emphasis is on workflow realism: turning abstract commitments into controls that any team member can understand and execute.

4.1 Why Principles Fail Without Process

Most organizations now have AI principles. They commit to fairness, transparency, accountability, privacy, safety, and human oversight. These principles are often borrowed from frameworks such as the OECD AI Principles, the EU Ethics Guidelines, or industry-specific standards. The problem is not the principles themselves. The problem is that principles are written in the language of values, while teams operate in the language of workflows, code, and configurations.

The implementation gap shows up in three ways. First, **ambiguity**: terms like "fairness" and "transparency" mean different things to different people. Without operational definitions, teams either ignore the principle or interpret it in the easiest way possible. Second, **missing mechanisms**: there is no defined point in the workflow where fairness is assessed, transparency is documented, or explainability is tested. Principles float above the process rather than being embedded in it. Third, **lack of accountability**: no one is explicitly responsible for ensuring that principles are followed. Teams assume someone else is checking.

For IT managers, closing the implementation gap is a core responsibility. You are the link between policy and practice. You cannot rely on data scientists to self-police without tools and standards. You cannot rely on legal to review every decision without clear criteria. You must build a system that automatically activates principles as teams move through design, development, testing, and deployment. This chapter gives you that system.

4.2 Translating Principles into Operational Definitions

The first step in operationalizing principles is to turn them into operational definitions. An operational definition specifies what a principle means in measurable, observable terms. It answers: What does this look like in practice? How do we know when we have achieved it? How do we test for it?

Take the principle of **fairness**. At a high level, fairness means that AI systems do not systematically disadvantage groups based on protected characteristics like race, gender, age, or disability. Operationally, fairness means: the system has been tested for disparate impact across demographic groups, performance metrics (accuracy, false positive rate, false negative rate) are comparable across groups, no protected attributes are used as direct inputs unless legally justified, and any observed disparities have been reviewed and documented with mitigation steps. Now, a team has something concrete to work with.

For **transparency**, the operational definition might be: users are informed when they interact with an AI system, the system's purpose and capabilities are explained in plain language, the types of data used are disclosed, and users have access to a contact point for questions or complaints. For **explainability**, it might be: the system can produce a human-readable explanation of its outputs; the explanation identifies the key factors influencing a decision; technical documentation is available for auditors; and explanations are tested for accuracy and clarity.

For **privacy**, the operational definition includes: only the minimum necessary data is collected; data is anonymized or pseudonymized where possible; access controls are enforced based on role; data retention limits are defined and enforced; and user consent is obtained where required. For **accountability**, this means an owner is assigned to the system, decision logs are maintained, audit trails are accessible, and there is a defined process for handling complaints or appeals.

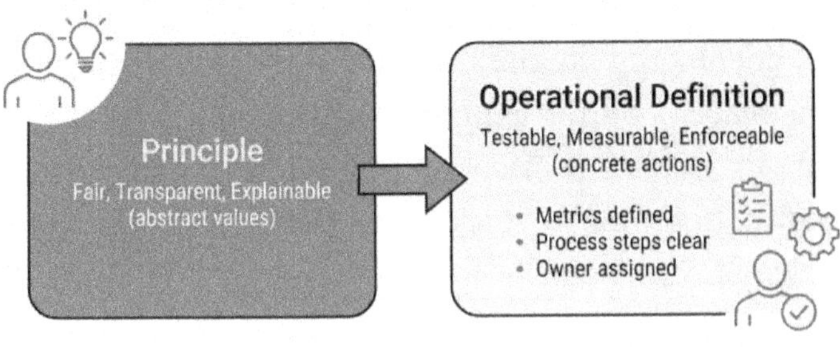

Turn values into actions teams can execute.

4.3 Embedding Fairness Checks into Model Releases

Fairness cannot be an afterthought. It must be built into the development lifecycle as a mandatory checkpoint. One of the most effective ways to do this is to create a **fairness gate** before production release. A fairness gate is a formal step in which the system is tested for bias and disparate impact, a cross-functional team reviews the results, and approval is required to proceed.

A fairness gate includes several components. First, **demographic parity testing**: if your use case involves people, collect demographic breakdowns (where legally permissible) and measure whether outcomes are comparable across groups. For example, in a hiring tool, are candidates from different demographic groups advancing at similar rates? In a loan approval tool, are approval rates comparable across lenders? Second, **performance equity testing**: measure model accuracy, false-positive rates, and false-negative rates across groups. A model that is highly accurate overall but systematically underperforms for one group is not fair.

Third, **feature audit**: review which features the model uses. Are any proxy variables indirectly encoding protected characteristics? For example, a ZIP code may correlate with race. If such features are present, document why they are necessary and what mitigation steps are in place. Fourth, **explanation review**: generate sample explanations for decisions affecting different demographic groups. Do the explanations make sense? Are they consistent? Do they reveal problematic patterns?

Finally, **mitigation and documentation**: if disparities are identified, teams must either mitigate them (rebalance training data, adjust thresholds, add fairness constraints to the model) or document a risk-acceptance decision with executive approval. The fairness gate does not demand perfect parity—context matters—but it does demand that disparities are known, understood, and addressed. No system passes the gate without completing this analysis.

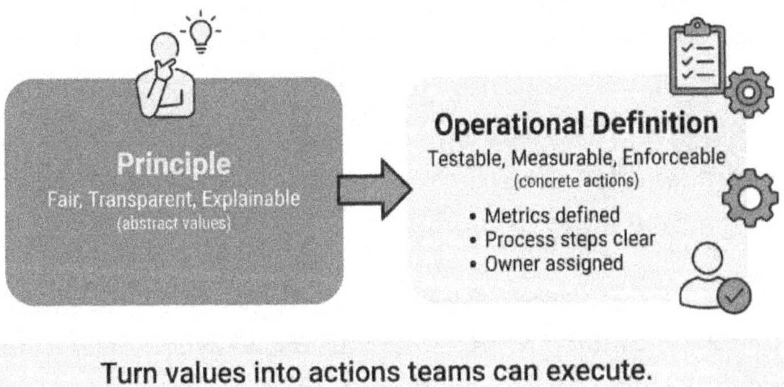

4.4 Transparency: Making AI Behavior Visible

Transparency is about making AI systems understandable and observable to stakeholders: users, auditors, regulators, and internal teams. Operationalizing transparency means building mechanisms that reveal what the system is doing, why it is doing it, and how to question or challenge it.

Start with **user-facing transparency**. When users interact with an AI system, they should know it is an AI system. Use clear labels: "An AI assistant generated this response," or "This recommendation is based on algorithmic analysis." Provide context: "We use your purchase history and browsing behavior to make recommendations. You can adjust your preferences here." Offer contact options: "If you have questions or concerns about this system, contact us at [email/phone]." This level of transparency builds trust and reduces surprise.

Next, implement **decision transparency** for high-impact use cases. If the AI system makes or influences decisions—hiring, lending, triage, eligibility—users should receive an explanation. The explanation does not need to be a technical breakdown of the model. It should be plain-language and actionable: "Your application was declined because your debt-to-income ratio exceeds our threshold. Here is what you can do to improve your chances in the future." For complex models, provide tiered explanations: a simple summary for most users, a detailed breakdown for those who request it, and full technical documentation for auditors.

Audit transparency is equally important. Maintain logs of key system events, including which data was used, which decisions were made, which explanations were provided, and which overrides occurred. These logs must be accessible to internal audit teams and, where required, to regulators. Structure logs so they are queryable and analyzable. A pile of unstructured logs is not transparent. Finally, publish a **model card or system card** for each AI system: a one- to two-page document describing the system's purpose, training data, performance metrics, known limitations, and intended use cases. Model cards are now considered best practice and are required by some regulations.

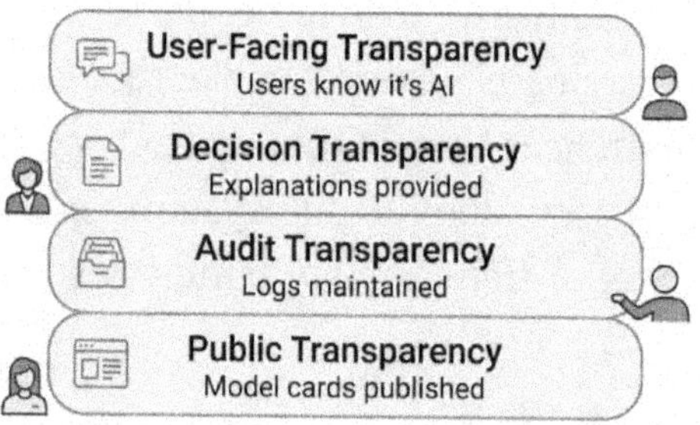

4.5 Explainability: From Black Box to Glass Box

Explainability is closely related to transparency but focuses specifically on the ability to understand why a system produced a particular output. For many stakeholders—users, auditors, and domain experts—explainability is essential for trust, accountability, and error correction. Operationalizing explainability means choosing models and architectures that support explanation, implementing explanation generation within the system, and testing explanations for accuracy and usefulness.

Explainability comes in two forms: **global** and **local**. Global explainability provides insight into how the model functions as a whole, including which features it prioritizes, the patterns it detects, and the decision boundaries it establishes. On the other hand, local explainability focuses on the reasoning behind a particular prediction for an individual instance. Incorporating both approaches is essential for a comprehensive operationalization strategy.

For global explainability, use methods such as feature importance analysis to identify which variables most strongly influence model predictions. Tree-based models facilitate this analysis conveniently, whereas for deep learning models, techniques like SHAP (SHapley Additive exPlanations) or attention visualizations are recommended. It is essential to document the overall model behavior in the model card. For local explainability, provide decision-level explanations using approaches such as rule-based surrogates, counterfactual reasoning (e.g., "If your income had been $X higher, you would have been approved"), or saliency maps when working with image models.

Explainability also requires **testing**. Generate explanations for a sample of decisions and have domain experts review them. Do the explanations make sense? Are they consistent? Do they align with organizational policy? If the model says "denied due to credit score," but the credit score was actually acceptable, the explanation is wrong or misleading. Fix the explanation logic or the model itself. Finally, make explanations **accessible**. For user-facing systems, explanations should be embedded in the interface, not buried in technical documentation. For audit systems, explanations should be queryable and exportable.

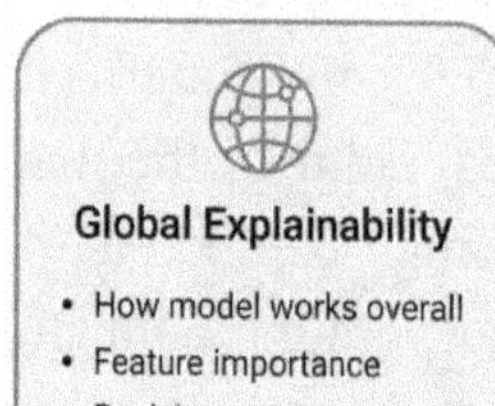

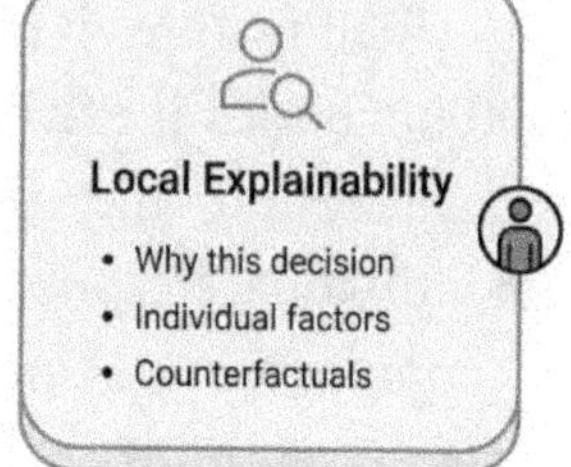

4.6 Data Minimization and Access Control

Privacy is a foundational principle, but it requires concrete technical and procedural practices to be effective. Operationalizing privacy means embedding data minimization, access control, anonymization, and retention policies into every stage of the AI lifecycle.

Data minimization is the practice of collecting and using only the data that is strictly necessary for the task. Before building a model, ask: What is the minimum set of features required to achieve acceptable performance? Can we exclude sensitive attributes entirely? Can we use aggregate statistics instead of individual records? Many teams collect as much data as possible "just in case," but this increases privacy risk and regulatory exposure. Start with the minimum and add more only if justified by a clear performance or business need.

Access control ensures that only authorized people and systems can access sensitive data. Use role-based access control (RBAC) or attribute-based access control (ABAC) to restrict data access. Log all access events. Regularly review access logs to detect anomalies. For AI systems, this means controlling which models can query which data sources, which applications can call which models, and which users can retrieve which outputs. Your IAM policies should be just as strict for AI as they are for your financial or HR systems.

Anonymization and pseudonymization reduce risk by removing or obfuscating personally identifiable information (PII). Where possible, train models on anonymized data. Use techniques like differential privacy to add noise that protects individual privacy while preserving aggregate patterns. For customer-facing systems, avoid storing PII in logs or explanation outputs.

Finally, **retention policies** define how long data is kept and when it is deleted. AI training data, model inputs, logs, and outputs should all have defined retention limits. Implement automated deletion processes to enforce these limits. Privacy is not a one-time decision; it is an ongoing operational discipline.

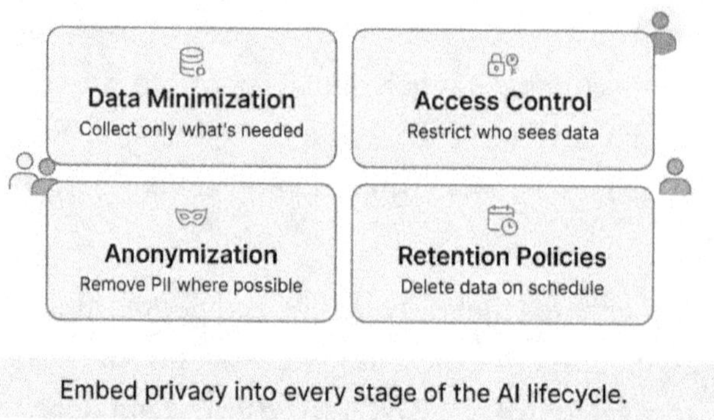

4.7 Accountability: Logging, Audit Trails & Overrides

Accountability means that when something goes wrong, you can determine what happened, who was responsible, and what corrective actions are needed. Operationalizing accountability requires robust logging, audit trails, and override mechanisms that capture key events and decisions throughout the AI system's lifecycle.

Logging is the foundation. Every AI system should log: when it was called, by whom, with what inputs, what outputs were generated, what explanations were provided, and what actions were taken. For high-risk systems, also log: confidence scores, which data sources were accessed, which model version was used, and whether any overrides occurred. Logs must be structured and timestamped to enable querying, analysis, and presentation to auditors. Treat AI system logs with the same rigor as transaction logs in financial systems.

Audit trails extend logging to include the full history of a system's development and deployment. Document: who proposed the use case, who approved it, what testing was conducted, what risks were identified, what mitigation steps were taken, when the system was deployed, and when it was updated or retired. This creates an evidence trail that auditors, regulators, or internal investigators can follow. Many regulatory frameworks explicitly require audit trails for high-risk AI systems.

Override mechanisms allow humans to intervene when AI systems make errors or when judgment is required. Every automated decision should have a clear path for human review and override. Log all overrides: who made them, when, and why. Analyze override patterns: frequent overrides in a particular area may indicate a model deficiency. Overrides are not failures; they are a safety valve. But they must be visible, logged, and reviewed. Finally, establish a **feedback loop**: when overrides occur, feed that information back to the model development team so future versions can improve.

Accountability Through Logging and Overrides

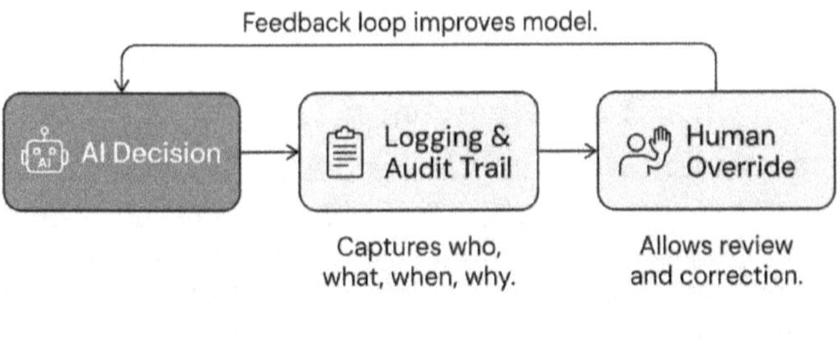

Make every decision traceable and reversible.

4.8 Aligning AI with your Ethics Framework

Many organizations already have a code of conduct, ethics framework, or values statement. AI should not exist in a separate silo. It should be explicitly tied to these existing commitments. This connection reinforces that AI is subject to the same organizational standards as any other business process and provides a familiar frame of reference for stakeholders.

Start by **mapping AI principles to your code of conduct**. If your code of conduct includes commitments to non-discrimination, integrity, respect, and transparency, show how your AI principles operationalize those commitments. For example, the fairness principle directly supports your non-discrimination commitment. The transparency principle supports integrity. This mapping makes AI feel less like a new, foreign domain and more like an extension of existing values.

Next, **incorporate AI into ethics training**. Many organizations provide regular ethics and compliance training to employees. Add modules on AI: what AI is, how it is used in your organization, what risks it introduces, and what responsibilities employees have. Make it clear that employees who interact with AI systems are expected to flag concerns, report anomalies, and escalate potential violations just as they would with any other ethics issue.

Establish a **clear reporting channel** for AI ethics concerns. Employees, customers, or partners should be able to report concerns about AI behavior without fear of retaliation. This might be an existing ethics hotline, an ombudsman, or a dedicated AI ethics contact. Publicize this channel and ensure reports are investigated and resolved promptly. Finally, **include AI governance in performance reviews and incentives**. If responsible AI is a stated priority, it should be reflected in how leaders and teams are evaluated. This signals that AI principles are not just words but expectations with real consequences.

Integrating AI into Organizational Ethics

4.9 Compliance Audits: From Checkbox to Culture

Compliance audits are often seen as bureaucratic checkboxes. In a mature AI governance program, they should be opportunities to surface issues, drive improvement, and reinforce culture. Operationalizing compliance means establishing a regular audit cadence, defining clear audit criteria, involving the right stakeholders, and acting on findings.

Start with **audit scope and frequency**. Define which AI systems are subject to audit and how often. High-risk systems (customer-facing, decision-making, sensitive data) should be audited at least annually, and more frequently if regulations require. Lower-risk systems might be audited every two years or spot-checked. Audits should cover: adherence to governance policies, fairness and bias testing, transparency and explainability mechanisms, privacy and data handling, logging and audit trails, incident response readiness, and documentation completeness.

Audit criteria should be specific and measurable. Instead of "Is the system fair?" ask: "Has the system been tested for demographic parity in the last 12 months? Are test results documented? Were disparities identified? If so, were mitigation steps taken or risks accepted with executive approval?" This turns audits from subjective judgment calls into objective assessments. Develop a **compliance audit checklist** and publish it so teams know what they will be evaluated against.

Involve cross-functional auditors: legal, compliance, security, domain experts, and technical reviewers. Each brings a different lens. Legal checks regulatory alignment. Security checks, access controls, and data handling. Domain experts check whether the system's behavior makes sense in context. Technical reviewers check whether the implementation matches the design. Finally, **act on audit findings**. If an audit reveals gaps, create corrective action plans with clear owners and deadlines. Track these plans to completion. Share audit summaries (anonymized if necessary) with leadership and the AI Working Group to drive systemic improvements.

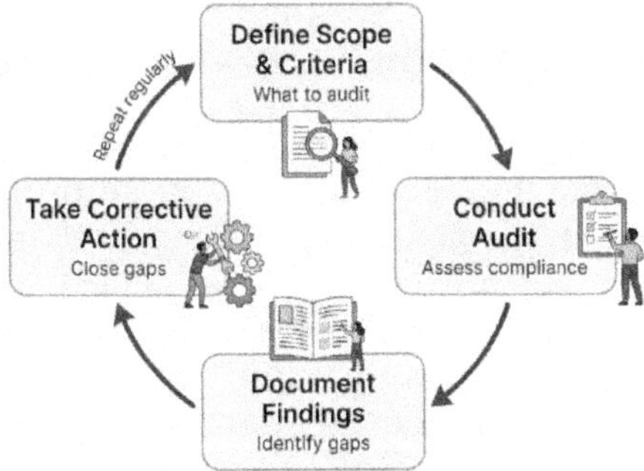

4.10   Manager's Playbook: Operationalizing Principles

You have seen the components: operational definitions, fairness gates, transparency mechanisms, explainability, privacy practices, accountability logging, ethics integration, and compliance audits. Now, how do you actually build this into your organization? Here is a practical, manager-ready playbook.

- **Step 1: Audit your current principles.** Pull up your organization's AI principles document. For each principle, ask: Is there an operational definition? Is there a process that enforces it? Is there a person or team responsible? Identify gaps. Most organizations will find that principles exist but operationalization is missing.

- **Step 2: Create operational definitions for each principle.** Use the frameworks in this chapter as templates. Write 3–5 bullet points per principle that define what it means in practice. Share these definitions with your AI Working Group and iterate until they are clear and actionable.

- **Step 3: Identify key checkpoints in your AI lifecycle.** Map your current process: ideation, design, data preparation, model training, testing, deployment, monitoring. Identify where each principle should be checked. For example, fairness might be checked at data preparation (balanced datasets), testing (bias testing), and monitoring (ongoing performance equity). Transparency might be checked at design (user interface includes disclosures) and deployment (model card published).

- **Step 4: Build checklists and templates.** For each checkpoint, create a simple checklist or form. For example, the fairness gate checklist might include: demographic parity tested (yes/no), performance equity tested (yes/no), feature audit completed (yes/no), disparities documented (yes/no), mitigation or risk acceptance approved (yes/no). Make these checklists mandatory and tracked.

- **Step 5: Train your teams.** Conduct workshops on what each principle means, how to use the checklists, and why it matters. Use real examples from your organization. Role-play scenarios where principles conflict or where judgment is required.

- **Step 6: Pilot the process.** Select 2–3 upcoming AI projects and run them through your new operationalized process. Observe where it works well and where it creates friction. Iterate. Simplify where possible. Clarify where teams are confused.

- **Step 7: Roll out and enforce.** Once the pilot is successful, roll out the process to all AI projects. Make adherence to the process a condition for production deployment. Conduct regular audits to ensure compliance. Celebrate teams that exemplify responsible practices.

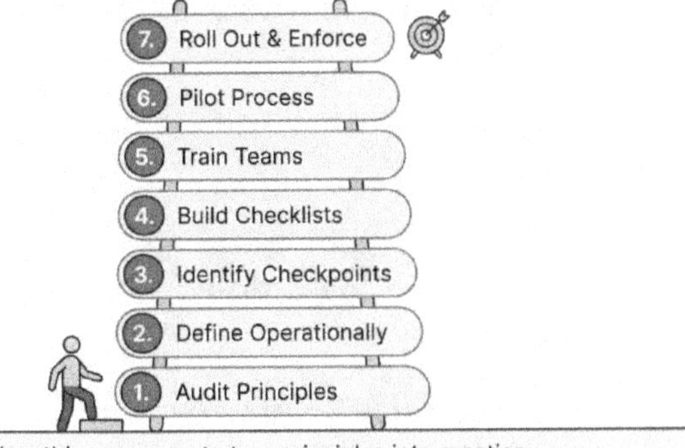

4.11   Common Challenges in Operationalizing Principles

Even with clear processes, you will encounter challenges. Here are five common obstacles and how to address them.

- **Challenge 1: Resistance from technical teams.** Data scientists and engineers may view operationalization as bureaucratic overhead that slows them down. **Solution:** Frame operationalization as risk reduction that protects them. Show how checklists and gates prevent late-stage surprises and rework. Involve technical leads in designing the processes so they feel ownership, not imposition.

- **Challenge 2: Conflicting principles.** Sometimes transparency conflicts with privacy, or fairness conflicts with accuracy. **Solution:** Acknowledge that principles involve trade-offs. Establish a decision-making framework for resolving conflicts: escalate to governance, document the trade-off and rationale, seek input from legal and ethics advisors, and make the decision explicit rather than implicit.

- **Challenge 3: Lack of tools.** Teams want to comply but do not have the tools to test for fairness, generate explanations, or anonymize data. **Solution:** Invest in tooling. Identify or build libraries for bias testing, explanation generation, and privacy-preserving techniques. Make these tools easy to integrate into existing workflows. Provide training and documentation.

- **Challenge 4: Insufficient expertise.** Not every team has a fairness expert or a privacy specialist. **Solution:** Create centers of excellence or embed experts in the AI Working Group who can be consulted. Provide training modules and office hours. Develop self-service resources: playbooks, decision trees, FAQs.

- **Challenge 5: Static processes in a dynamic environment.** AI capabilities, risks, and regulations evolve rapidly. Processes designed today may be obsolete in six months. **Solution:** Treat operationalization as a living practice. Schedule quarterly reviews of your operational definitions, checklists, and gates. Update them as you learn and as the environment changes. Communicate updates clearly to teams.

4.12   Embedding Controls Without Breaking Speed

One of the most important concepts in operationalizing principles is workflow realism. Controls that are too heavyweight or disconnected from how teams actually work will be ignored or bypassed. Controls that are embedded, automated, and frictionless will be adopted naturally. Your goal is to design guardrails that feel like part of the workflow, not external obstacles.

Embed controls into existing tools. If your teams use a model registry, add fields for fairness test results, model card links, and approval status. If they use a project management tool, add stages for governance review. If they use CI/CD pipelines, add automated checks for bias, privacy, and explainability. The more integrated the controls, the less friction they create.

Automate where possible. Manual checklists are better than nothing, but automated checks are better still. Automate bias testing by running standard fairness metrics on every model before deployment. Automate privacy checks by scanning code and data access patterns for PII exposure. Automate audit trail generation by instrumenting your systems to log key events automatically. Automation reduces burden and ensures consistency.

Make controls visible. Teams should see progress through the governance process, not wonder where their project is stuck. Use dashboards, status indicators, and clear communication. When a fairness gate is passed, celebrate it. When a gap is found, make it visible along with a clear path to remediation. Transparency about the process builds trust in the process.

Balance rigor with pragmatism. Not every use case needs the same level of scrutiny. Use your risk tiers from Chapter 2 to calibrate control intensity. Low-risk use cases get lightweight checklists. High-risk use cases get full audits and cross-functional review. This proportionality keeps the overall governance system sustainable.

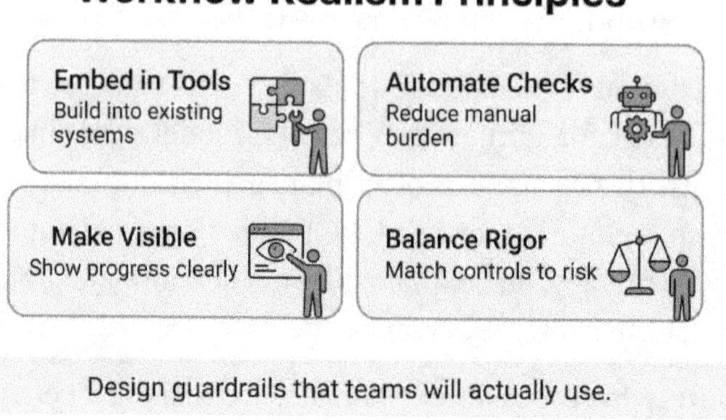

4.13 Case: Operationalizing Fairness in a Hiring Tool

To make these concepts concrete, consider a case example: operationalizing fairness in an AI-powered hiring tool. The organization has a principle that AI must be fair and non-discriminatory. Here is how they operationalize it.

Operational definition: The hiring tool must produce demographically comparable outcomes across protected groups, measure performance equity (false positive and false negative rates), exclude direct use of protected attributes, and document any observed disparities with mitigation steps.

Fairness gate: Before deployment, the tool is tested on historical hiring data broken down by gender, race, age, and disability status (where data is available and legal). Metrics include: selection rate parity (are candidates from different groups selected at similar rates?), interview advancement parity (do candidates advance to interviews at similar rates?), and final offer parity (are offers extended at similar rates?).

Results show that the tool selects women at a 5 percent lower rate than men. The team investigates and finds that the model heavily weights years of experience, and women in the candidate pool have slightly fewer years on average due to career breaks. The team decides to adjust the model to place less weight on raw years of experience and more on demonstrated skills and achievements. After retraining, the selection rate gap drops to 1 percent, within acceptable limits.

Documentation: The team documents the original disparity, the investigation, the mitigation steps, and the final test results. This documentation is reviewed by the AI Working Group and approved. The tool is deployed with ongoing monitoring: quarterly fairness audits to ensure performance remains equitable. If disparities reappear, the system is paused and re-evaluated. This is operationalized fairness: not a vague aspiration, but a concrete, repeatable process with checkpoints, metrics, and accountability.

4.14   AI Reality Check

Before we close, a reality check on common myths about operationalizing principles:

- **Myth 1: "Operationalizing principles is a one-time project."**
  Reality: It is an ongoing practice. As AI systems evolve, risks change, and regulations update, your operational definitions and processes must evolve too.
- **Myth 2: "If we have principles, we are compliant."**
  Reality: Principles without processes provide no protection. Regulators and auditors look for evidence of implementation, not just statements of intent.
- **Myth 3: "Fairness and explainability are too hard to measure."**
  Reality: They require effort, but they are measurable. There are established frameworks, tools, and metrics. The challenge is organizational will, not technical impossibility.
- **Myth 4: "Operationalizing principles will slow us down."**
  Reality: Poor operationalization will slow you down. Good operationalization, embedded and automated, creates clarity and confidence that speeds you up.

Use this reality check to reset expectations with leadership and teams. Operationalizing principles is hard work, but it is necessary work, and it is achievable with the right structure, tools, and commitment.

4.15 Summary: From Values to Action

This chapter has shown you how to bridge the implementation gap: the space between high-level AI principles and day-to-day operational reality. You have learned to translate principles into operational definitions, embed fairness checks into release gates, implement transparency mechanisms at multiple layers, operationalize explainability through global and local techniques, enforce privacy through data minimization and access control, build accountability through logging and audit trails, tie AI to your existing ethics framework, and establish compliance audits that drive continuous improvement.

The key to success is workflow realism: designing controls that are embedded, automated, visible, and proportional. Operationalizing principles is not a one-time project. It is an ongoing discipline that evolves with your AI portfolio, your organizational maturity, and the external environment. Your role as an IT manager is to champion this discipline, provide the tools and training teams need, and hold the line when shortcuts are proposed. When principles are operationalized effectively, they stop being abstract aspirations and become tangible competitive advantages: trust, accountability, and the confidence to scale AI safely.

5 Technical Guardrails for Responsible AI

Opening Scenario: When Technical Controls Are Missing

Your customer service chatbot went live three weeks ago. It handled routine inquiries beautifully, reduced call volume, and earned positive feedback. Then someone asked it about a competitor's product. The bot not only provided detailed information about the competitor but also linked to their website. Another user discovered they could manipulate the bot into revealing internal policy details by framing requests as "training exercises." A third user reported that the bot occasionally generated responses containing what appeared to be fragments of other customers' conversations. None of these issues showed up in pre-launch testing. They emerged because the system lacked technical guardrails: input filters, output validators, data isolation controls, and adversarial testing. Your governance policies were clear, but the technical enforcement was absent. This is the gap that technical guardrails fill.

Technical guardrails are the technologies, architectures, and code-level controls that enforce your policies and principles at the system level. While governance defines what should happen, technical guardrails ensure it actually happens. This chapter gives managers a practical understanding of the technical tools available for implementing guardrails: dataset validation, prompt filtering, output screening, red-team simulation, automated bias detection, and interpretability frameworks. The perspective is not that of a coder, but of a governance leader who needs to know what to ask for, how to measure effectiveness, and how to ensure that guardrails are functioning in production, not just existing on paper.

5.1 The Technical Layer: Where Policy Meets Code

Technical guardrails are the mechanisms that translate policy into enforceable constraints. If your policy says "do not use sensitive customer data without authorization," the technical guardrail is the access control system that prevents unauthorized queries. If your policy says "do not generate harmful or inappropriate content," the technical guardrail is the output filter that blocks such content. If your policy says "log all decisions for audit," the technical guardrail is the instrumentation that captures and stores those logs.

For IT managers, this layer is familiar territory. You already manage firewalls, authentication systems, encryption, and monitoring tools. AI systems require analogous controls, adapted to the unique properties of probabilistic models. Unlike traditional software, AI systems do not follow deterministic rules. They learn patterns from data and generate outputs based on statistical likelihoods. This means that guardrails must account for uncertainty, drift, and emergent behavior. You cannot simply review the code and know exactly what the system will do. You need dynamic, runtime controls that continuously monitor and constrain behavior.

Technical guardrails operate at multiple points in the AI lifecycle: during data preparation (ensuring datasets are clean, representative, and privacy-preserving), during model development (incorporating fairness constraints, interpretability hooks, and robustness checks), at inference time (filtering inputs, validating outputs, enforcing rate limits), and during operation (monitoring performance, detecting drift, logging interactions). This chapter walks through each category, providing a manager's lens on what these controls do, why they matter, and how to verify they are working.

5.2 Source-Level Data Quality Checks

The foundation of any AI system is its data. If the training data is biased, incomplete, or corrupted, no amount of downstream guardrails will fully compensate. Dataset validation is the practice of systematically checking data for quality, representativeness, privacy compliance, and integrity before it is used to train or fine-tune models.

Start with **data quality checks**. These are automated tests that verify basic integrity: Are there missing values? Are there duplicate records? Are numerical fields within expected ranges? Are categorical fields consistent? Many data quality issues are mundane but consequential. A missing value might cause a model to treat absence as zero, distorting predictions. A duplicated record might cause the model to overweight certain patterns. Establish thresholds: if more than 5% of records are missing critical fields, flag the dataset for review before use.

Next, implement **representativeness checks**. If your AI system will serve a diverse population, the training data should reflect that diversity. For example, if you are building a hiring tool, does your training data include candidates from different demographic groups, educational backgrounds, and career paths? If it is skewed toward one group, the model will perform poorly for others. Representativeness is not just about demographics; it is also about edge cases, rare events, and evolving patterns. Document the known limitations of your datasets and assess whether they pose unacceptable risks to your use case.

Privacy compliance checks ensure that datasets do not contain data that should not be there. Scan for personally identifiable information (PII) that should have been anonymized or removed. Check that data access permissions were properly obtained and documented. Verify that data retention policies are being followed: training data should not include records that should have been deleted. Finally, implement **integrity checks** to detect tampering or corruption. Use checksums, version control, and access logs to ensure that datasets have not been altered without authorization. Data provenance—knowing where data came from and how it was transformed—is essential for auditability.

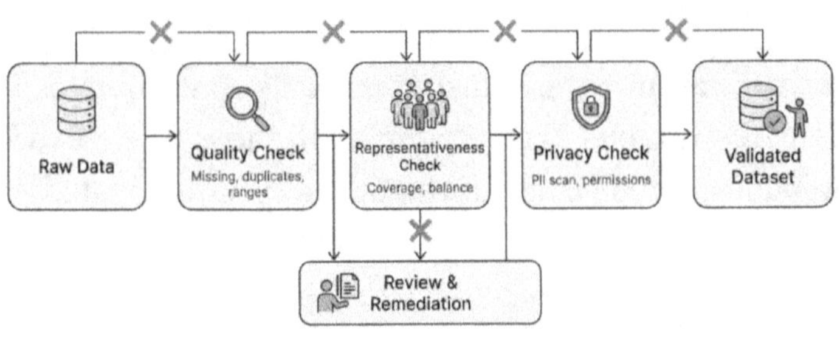

Validate data before it enters your AI pipeline.

5.3 Prompt Filtering and Input Validation

For generative AI systems, especially those using large language models, the prompt is the user's input that directs the model's behavior. Prompt filtering and input validation are guardrails that prevent users from manipulating the system into unsafe, inappropriate, or policy-violating outputs. These controls sit at the entry point of the system, analyzing and modifying inputs before they reach the model.

Prompt filtering involves scanning user inputs for patterns that indicate adversarial intent or policy violations. Common filtering rules include: blocking prompts that attempt to override system instructions (jailbreaking), blocking prompts that request harmful or illegal content (violence, explicit material, instructions for illegal activity), blocking prompts that attempt to extract sensitive information (social engineering), and blocking prompts that contain abusive or harassing language. Filters can be rule-based (pattern matching, keyword lists) or model-based (using a classifier to detect adversarial prompts).

Input validation ensures that user inputs conform to expected formats and constraints. For example, if the system expects a customer inquiry, validate that the input is text of reasonable length, does not contain executable code or scripts, and does not include patterns associated with injection attacks. For structured inputs (forms, APIs), enforce schema validation: check that required fields are present, that data types are correct, and that values fall within acceptable ranges.

One challenge is **balancing safety and usability**. Overly aggressive filtering will block legitimate user inputs, creating frustration. For example, a filter that blocks all mentions of "weapons" might also block a history teacher asking about medieval weaponry. The solution is to use tiered filtering: high-confidence threats are blocked immediately, medium-confidence inputs are flagged for human review or given warnings, and low-confidence inputs are allowed but logged for monitoring. Continuously review false positives and false negatives to refine your filters.

Finally, consider **prompt templating**. Instead of allowing free-form user inputs directly to the model, use structured templates that fill in user-provided values within a predefined format. For example, a customer support bot might use the template: "The user asked: [USER_INPUT]. Provide a helpful response based on our knowledge base, without disclosing internal policies." This constrains the model's behavior and reduces the attack surface.

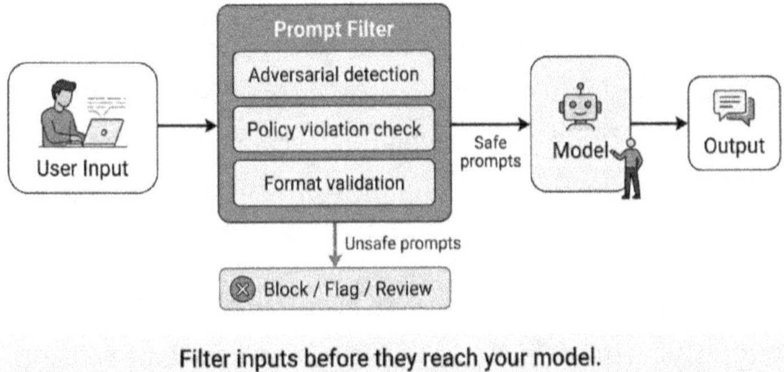

Filter inputs before they reach your model.

5.4 Output Screening and Content Moderation

Even with input filtering, generative models can produce outputs that violate policies or are inappropriate. Output screening and content moderation are guardrails that analyze and filter the model's responses before they reach users. These controls provide a final safety check, ensuring that what leaves the system aligns with organizational standards.

Output screening uses classifiers or rule-based systems to evaluate generated content. Common screening dimensions include: harmfulness (does the output contain violent, explicit, or dangerous content?), toxicity (is the output abusive, hateful, or harassing?), factual accuracy (does the output contain known misinformation or unsupported claims?), policy compliance (does the output violate organizational policies, such as disclosing confidential information?), and coherence (is the output nonsensical or off-topic?). Outputs that fail screening can be blocked, replaced with a safe default response, or flagged for human review.

Content moderation strategies vary by risk tolerance and use case. In high-risk scenarios (customer-facing, legal, medical), use strict screening with low tolerance for false negatives (missing harmful content). Accept that this will create false positives (blocking safe content), and provide a mechanism for users to request human review. In lower-risk scenarios (internal tools, advisory only), use lighter screening to maintain fluency and flexibility, but log all outputs for periodic review.

For **hallucination detection**, a particularly challenging issue with generative models, implement confidence scoring and grounding checks. If the model generates a factual claim, check whether it is supported by the knowledge base or retrieval context provided. If the model's confidence is low or the claim cannot be verified, either suppress the output or add a disclaimer: "This information may not be accurate. Please verify independently." For domains like healthcare or legal advice, this guardrail is critical.

Finally, implement **redaction and masking** for sensitive information. Screen outputs for PII, confidential data, or proprietary information. If detected, redact it before displaying to users. For example, if the model accidentally generates a social security number or an internal document reference, replace it with "[REDACTED]" or a generic placeholder. Log these events for investigation: why is the model generating sensitive information, and how can training or fine-tuning prevent it?

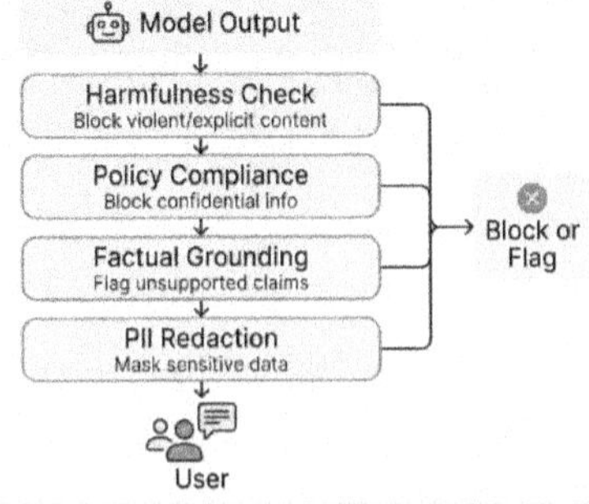

Screen outputs before they reach users.

5.5 Red-Team Simulation and Adversarial Testing

Guardrails are only effective if they can withstand real-world adversarial pressure. Red-team simulation and adversarial testing are practices where you deliberately try to break your own system, exposing weaknesses before bad actors do. For AI systems, this means attempting to manipulate the model into unsafe behavior, bypass filters, leak data, or produce policy-violating outputs.

Red teaming involves assembling a team (internal or external) whose job is to attack the AI system using creative, adversarial techniques. Red teamers might try prompt injection (embedding instructions that override system prompts), social engineering (tricking the model into revealing information by posing as an authorized user), edge case exploitation (finding unusual inputs that confuse the model), and multi-turn manipulation (building trust over several interactions before making an adversarial request). The goal is not to prove the system is perfect, but to discover vulnerabilities so they can be mitigated.

Adversarial testing complements red teaming with automated approaches. Use adversarial example generators to create inputs designed to fool the model: for text models, this might involve paraphrasing, obfuscation, or using synonyms to evade keyword filters; for image models, this might involve pixel perturbations or adversarial patches. Test boundary conditions: what happens if a user submits an extremely long prompt? What if they repeat the same word 1,000 times? What if they include special characters, code snippets, or foreign language text?

Document all findings in a **red-team report**: what vulnerabilities were discovered, what their potential impact is, and what mitigation steps are recommended. Prioritize findings by severity (critical, high, medium, low) and likelihood. Treat red-team findings like security vulnerabilities: track them, remediate them, and verify fixes. Conduct red-team exercises regularly—at least annually for high-risk systems—and after significant system changes (new model, new feature, new data source).

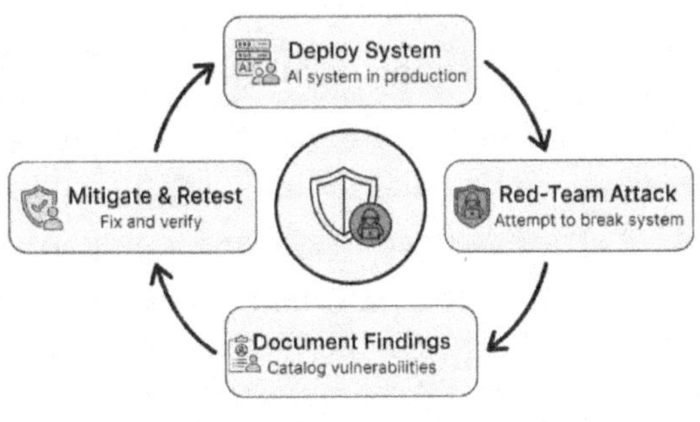

Test your guardrails by trying to break them.

5.6 Automated Bias Detection and Fairness Testing

Bias in AI systems is a persistent risk. Automated bias detection tools allow you to continuously test models for fairness, rather than relying solely on manual reviews. These tools analyze model predictions across demographic groups, flagging disparities that may indicate bias. For managers, the key is to understand what these tools measure, how to interpret results, and when to escalate findings.

Common fairness metrics include demographic parity (do different groups receive positive outcomes at similar rates?), equalized odds (are false positive and false negative rates similar across groups?), and predictive parity (is accuracy comparable across groups?). No single metric captures all dimensions of fairness, so use multiple metrics and interpret them in context. For example, demographic parity might not be appropriate if base rates differ across groups (e.g., disease prevalence in medical diagnosis). Equalized odds might be more relevant.

Automated bias detection tools typically work by: ingesting model predictions and ground truth labels, segmenting data by demographic attributes (if available and legally permissible), computing fairness metrics for each segment, comparing metrics across segments to identify disparities, and generating reports that flag statistically significant differences. Some tools also provide explanations: which features are contributing most to observed disparities?

When disparities are detected, do not assume the model is inherently biased. Investigate root causes: Is the training data skewed? Are certain features acting as proxies for protected attributes? Is the model underfitting for certain groups due to insufficient representation? Is there a legitimate reason for the disparity (e.g., different risk profiles in a fraud detection system)? Use this analysis to decide on mitigation: rebalance training data, remove or adjust problematic features, apply fairness constraints during training, or adjust decision thresholds for different groups.

Automate bias testing in your CI/CD pipeline. Every time a model is retrained or updated, run automated fairness tests. If metrics fall outside acceptable bounds, block deployment and route the issue for review. This ensures that fairness is not a one-time check, but an ongoing commitment.

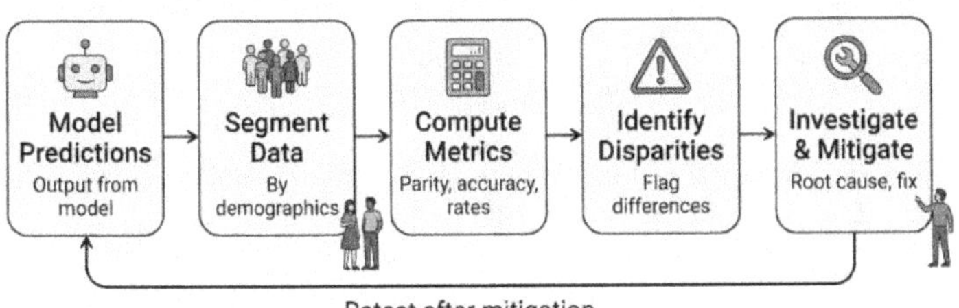

5.7 Interpretability and Explainability Tools

Interpretability and explainability are technical capabilities that allow you to understand and communicate how a model makes decisions. For managers, these tools are essential for building trust, satisfying regulatory requirements, and debugging problematic behavior. While Chapter 3 covered the policy side of explainability, this section focuses on the technical tools that generate explanations.

Global interpretability tools reveal overall model behavior. Common techniques include feature importance analysis (which features have the most influence on predictions?), partial dependence plots (how does changing a single feature affect predictions?), and rule extraction (can the model's logic be approximated with human-readable rules?). For example, a feature importance analysis might reveal that "credit score" is the most influential factor in a loan approval model, while "ZIP code" is unexpectedly influential—a red flag for potential proxy bias.

Local interpretability tools explain individual predictions. Techniques include SHAP (SHapley Additive exPlanations), which assigns each feature a contribution value for a specific prediction, LIME (Local Interpretable Model-agnostic Explanations), which approximates the model's behavior locally with a simpler model, and counterfactual explanations, which show what changes to inputs would have led to a different outcome. For example, a SHAP explanation might show: "This loan application was denied because credit score contributed -50 points, debt-to-income ratio contributed -30 points, and employment history contributed +10 points."

For **deep learning models**, interpretability is more challenging. Use attention visualization (for transformer models, show which parts of the input the model focused on), saliency maps (for image models, highlight regions that influenced the prediction), and layer-wise relevance propagation. These techniques provide partial insight but are not as definitive as interpretability for simpler models. In high-risk use cases, this may justify choosing simpler, more interpretable models over complex deep learning models.

Integrate interpretability tools into your production systems so explanations are generated automatically. For customer-facing systems, surface explanations in the user interface. For auditors, provide dashboards that allow querying explanations for specific decisions. Test explanations for consistency and accuracy: do they align with what the model is actually doing, or are they post-hoc rationalizations that do not reflect true model logic?

Interpretability Techniques

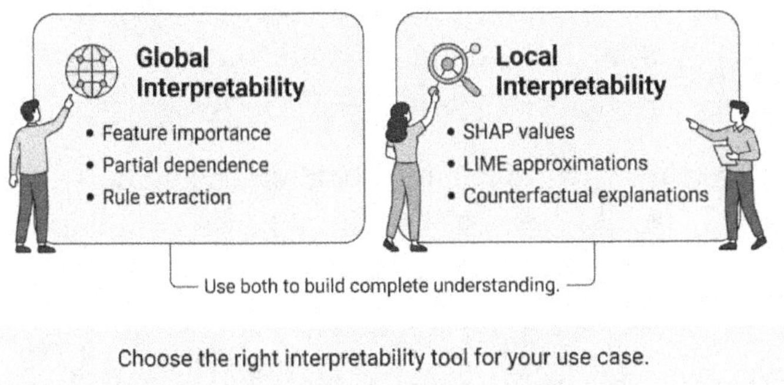

Choose the right interpretability tool for your use case.

5.8 Rate Limiting and Abuse Prevention

AI systems, especially those exposed via APIs or web interfaces, are vulnerable to abuse: users making excessive requests to overload the system, scraping data, or probing for vulnerabilities. Rate limiting and abuse prevention are technical guardrails that protect system availability, prevent data leakage, and deter adversarial behavior.

Rate limiting restricts the number of requests a user or client can make within a time window. For example, limit each user to 100 requests per hour, or limit each API key to 1,000 requests per day. Rate limits should be calibrated to legitimate use patterns: too low, and you frustrate real users; too high, and you fail to prevent abuse. Implement tiered rate limits: authenticated users get higher limits than anonymous users, premium users get higher limits than free users.

Abuse detection goes beyond simple rate limits by identifying suspicious patterns: users who repeatedly submit adversarial prompts, users who systematically probe for sensitive data, users whose queries suggest automated scraping, or users who exhibit sudden spikes in activity. Use heuristics and machine learning models to flag these patterns. When abuse is detected, escalate: increase rate limits, require additional authentication (CAPTCHA, MFA), or temporarily block the user pending investigation.

For **API security**, enforce authentication and authorization for all endpoints. Use API keys, OAuth tokens, or other secure mechanisms to identify clients. Log all API calls, including client identity, request content, response content, and timestamps. Monitor for anomalies: unusual access patterns, spikes in errors, or requests from unexpected geographic locations. Implement circuit breakers: if a downstream service (like a model endpoint) becomes overloaded, temporarily reject new requests to prevent cascading failures.

Finally, consider **cost controls**. Generative AI models can be expensive to run. Rate limiting prevents a single user from incurring excessive costs. Implement budget alerts: notify administrators if usage costs exceed thresholds. For internal users, track costs by team or project and provide transparency into who is consuming resources. This turns rate limiting from a purely defensive mechanism into a cost management tool.

Rate Limiting and Abuse Prevention

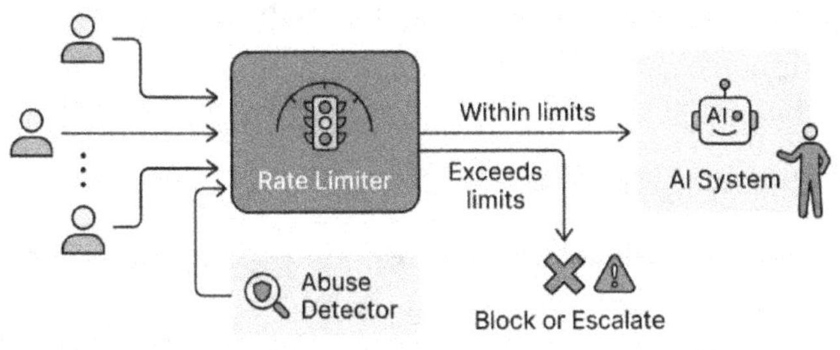

Protect your system from overload and adversarial abuse.

5.9 Access Control in Multi-Tenant Systems

Many AI systems serve multiple customers, business units, or user groups. Data isolation and access control ensure that one tenant cannot access another tenant's data, and that the model does not inadvertently leak information across boundaries. These guardrails are especially critical for customer-facing systems and platforms.

Data isolation means that training data, inference inputs, and system outputs are logically or physically separated by tenant. For example, if you offer an AI assistant to multiple enterprise customers, each customer's data must be isolated so that queries from Customer A do not retrieve or reference data from Customer B. Implement isolation at the data layer (separate databases or schemas per tenant), at the model layer (separate fine-tuned models per tenant or strict prompt segmentation), and at the logging layer (separate logs per tenant).

Access control enforces who can access which data and models. Use role-based access control (RBAC) to define permissions: which users can query which models, which users can view logs, which users can update configurations. For multi-tenant systems, access control must be tenant-aware: users from Tenant A should only see Tenant A's data, even if they have admin privileges within their tenant. Implement strict authentication and session management: validate user identity on every request, enforce short session timeouts, and log all access attempts.

Prompt and context management is a subtle but important guardrail. In generative AI systems, the model's context window can inadvertently mix information from different users or sessions. For example, if User 1's conversation is still in memory when User 2's query arrives, the model might reference User 1's data. Implement session isolation: each user session gets a clean context, and no cross-session data is retained. For retrieval-augmented generation (RAG) systems, ensure that retrieval queries are scoped to the user's authorized data only.

Finally, conduct **regular isolation audits**. Test whether it is possible to cross tenant boundaries through adversarial queries or configuration errors. Use penetration testing and red-team exercises specifically targeting data isolation. A single isolation breach can destroy trust and violate regulations. Treat data isolation as a critical security control, not an optional feature.

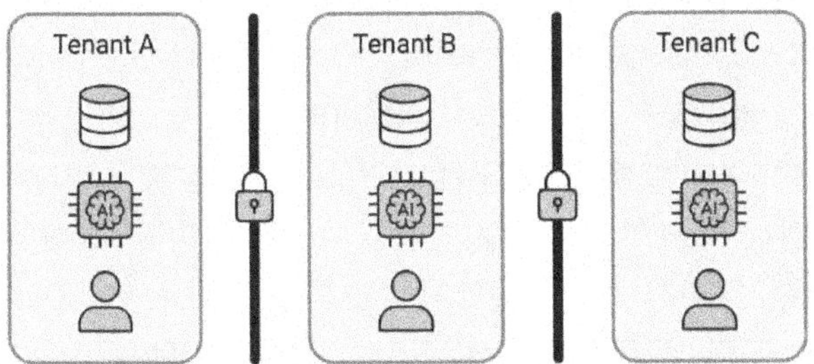

Each tenant's data, model, and logs are isolated.

Enforce strict isolation in multi-tenant AI systems.

5.10 Monitoring Instrumentation and Logging Arch.

Technical guardrails are only effective if you can observe whether they are functioning. Monitoring instrumentation and logging architecture provide the visibility needed to detect issues, investigate incidents, and demonstrate compliance. For managers, this means ensuring that your AI systems are as observable as your traditional IT systems.

Monitoring instrumentation involves embedding hooks throughout the AI system to capture key metrics and events. At the input layer, log: number of requests, sources of requests, prompt lengths, filtered prompts (how many were blocked?). At the model layer, log: inference latency, model version used, confidence scores, feature distributions. At the output layer, log: response lengths, screened outputs (how many were blocked?), user feedback signals (thumbs up/down, reports). At the system layer, log: CPU and memory usage, error rates, rate limit violations, authentication failures.

Logging architecture must be structured, queryable, and scalable. Use structured logging formats (JSON) rather than unstructured text. Include consistent fields in every log entry: timestamp, user ID, session ID, tenant ID, model ID, request ID, and event type. Store logs in a centralized system (log aggregator, data warehouse) that supports querying and analysis. Enforce log retention policies: retain high-priority logs (errors, security events, audit trails) longer than low-priority logs (routine operations).

For **compliance and audit**, logs must be tamper-proof and accessible. Use append-only logging systems or write logs to immutable storage. Implement access controls for logs: only authorized personnel (auditors, security team, governance team) can access them. Provide audit trails for log access itself: who viewed logs, when, and why. Many regulatory frameworks require specific log retention periods (e.g., 7 years for financial data). Document your retention policies and automate enforcement.

Finally, implement **alerting and dashboards**. Configure alerts for critical events: guardrail failures (e.g., output screening blocked a high volume of responses), performance degradation (e.g., response time exceeds threshold), security events (e.g., repeated authentication failures), or compliance violations (e.g., unauthorized data access). Create dashboards that visualize key metrics: request volume over time, filter hit rates, model performance by demographic group, error rates, and cost trends. Make these dashboards accessible to governance and operational teams.

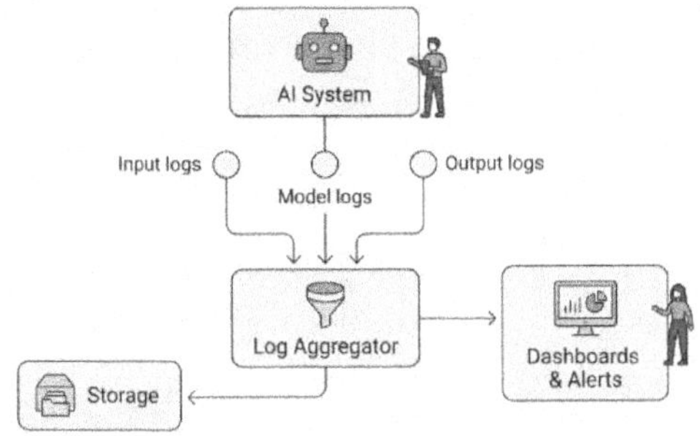

Build observability into every layer of your AI system.

5.11 Secure Model Deployment and Versioning

Models are intellectual property and critical infrastructure. Secure model deployment and versioning practices protect models from tampering, ensure reproducibility, and support rollback when issues arise. For managers, this is analogous to secure software deployment, but adapted for the unique properties of AI artifacts.

Model versioning means treating every trained model as a versioned artifact. Assign each model a unique identifier (e.g., semantic versioning: v1.2.3) and track metadata: training date, training data version, hyperparameters, performance metrics, who trained it, and who approved it for deployment. Store models in a **model registry**—a centralized repository that catalogs all models, their versions, and their lineage. The registry should support: search and discovery (which models are available?), access control (who can download or deploy models?), and audit trails (who deployed which model, when?).

Secure deployment means protecting models from unauthorized access or modification. Encrypt models at rest and in transit. Use signed containers or checksums to verify model integrity: when a model is loaded, verify that it has not been tampered with. Restrict deployment permissions: only authorized CI/CD pipelines or operators can deploy models to production. Log all deployment events: which model was deployed, to which environment, by whom, and when.

Rollback capability is essential. If a newly deployed model causes issues (performance degradation, fairness problems, user complaints), you must be able to quickly revert to the previous version. Implement blue-green deployments or canary deployments: deploy the new model alongside the old model, route a small percentage of traffic to the new model, monitor for issues, and gradually increase traffic if the model performs well. If issues arise, route all traffic back to the old model. Treat model deployment with the same rigor as critical software releases.

Finally, implement **model expiration policies**. Models trained on historical data may become stale as patterns shift. Set expiration dates or performance thresholds: if a model is more than X months old or if performance drops below Y threshold, trigger a retraining workflow. This prevents old, degraded models from lingering in production.

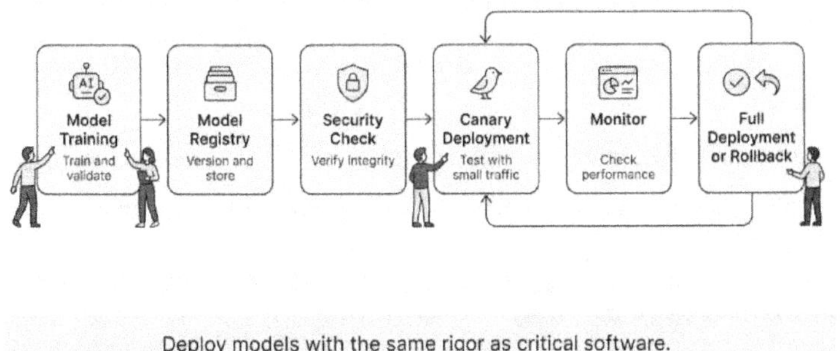

Deploy models with the same rigor as critical software.

5.12   Asking the Right Questions About AI Guardrails

You do not need to implement these technical guardrails yourself. Your technical teams will do the hands-on work. But you do need to know what to ask for, how to verify it is working, and how to hold teams accountable. Here is a practical playbook of questions to ask.

For dataset validation: Have we documented the provenance of all training data? Have we run quality checks (missing values, duplicates, outliers)? Have we assessed representativeness across key demographic groups? Have we scanned for PII and verified data access permissions? Are these checks automated and part of our data pipeline?

For prompt filtering: What types of adversarial prompts are we filtering for? Are filters rule-based, model-based, or both? How often are filters updated? What is our false positive rate (legitimate prompts blocked)? Are blocked prompts logged and reviewed?

For output screening: What dimensions of content are we screening (harmfulness, toxicity, policy violations, PII)? What is our threshold for blocking vs flagging outputs? Are screened outputs logged? Do we have a process for reviewing false positives and false negatives?

For red teaming: When was the last red-team exercise conducted? Who conducted it (internal or external)? What vulnerabilities were discovered? Have they been remediated? Are findings tracked like security vulnerabilities?

For bias detection: Are we testing models for fairness across demographic groups? Which fairness metrics are we using? Are tests automated in our CI/CD pipeline? What happens if a model fails fairness tests? Are disparities documented and approved?

For interpretability: Can we generate explanations for individual decisions? Are explanations tested for accuracy and consistency? Are they accessible to users, auditors, and domain experts? For deep learning models, what interpretability techniques are we using?

For rate limiting: What rate limits are in place? How were they calibrated? Are they tiered by user type? Do we monitor for abuse patterns? What happens when abuse is detected?

For data isolation: How do we isolate data between tenants or user groups? Are there separate databases, models, or strict prompt scoping? Have we tested for cross-tenant data leakage? When was the last isolation audit?

For monitoring: What metrics are we logging at input, model, output, and system layers? Are logs structured and queryable? What is our log retention policy? Are critical events alerted? Are dashboards accessible to governance teams?

For model deployment: Are all models versioned and stored in a registry? Are deployment permissions restricted? Do we verify model integrity before loading? Can we roll back to previous model versions? What is our model expiration policy?

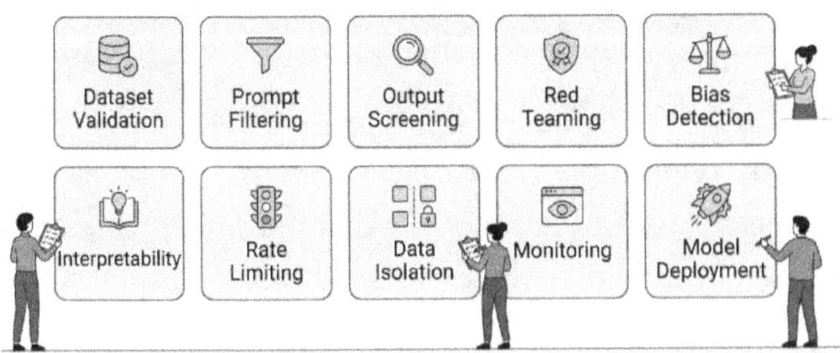

5.13 Measuring Guardrail Effectiveness

Having technical guardrails in place is not enough. You must measure their effectiveness continuously. Guardrails can degrade over time, become misconfigured, or fail to keep pace with evolving threats. Establish key performance indicators (KPIs) for each guardrail category and track them regularly.

For input filtering: Track filter hit rate (percentage of prompts flagged or blocked), false positive rate (legitimate prompts incorrectly blocked), false negative rate (adversarial prompts that bypassed filters), and time to update filters after new threat patterns are identified.

For output screening: Track screening hit rate (percentage of outputs flagged or blocked), false positive and false negative rates, redaction rate (percentage of outputs requiring PII redaction), and user complaint rate (are users reporting inappropriate outputs that screening missed?).

For bias detection: Track fairness metrics over time (are they stable or degrading?), time to detect bias (how long between model deployment and bias identification?), time to remediate (how long to address detected bias?), and number of use cases tested (what percentage of your AI portfolio has been assessed for fairness?).

For red teaming: Track the number of vulnerabilities discovered per exercise, severity distribution (critical, high, medium, low), time to remediate vulnerabilities, and retest success rate (percentage of vulnerabilities successfully closed).

For monitoring: Track log completeness (are all expected events being logged?), alert response time (how quickly do teams respond to alerts?), and mean time to detect incidents (MTTD) and mean time to resolve incidents (MTTR).

Establish baselines and targets. For example, a target might be: input filter false positive rate below 2 percent, output screening false negative rate below 0.5 percent, fairness test coverage at 100 percent of customer-facing models, red-team exercises at least annually, and MTTD for critical incidents under 15 minutes. Review these KPIs in your AI Working Group meetings. If metrics are degrading, investigate and take corrective action.

5.14   AI Reality Check

Before we close, let's address common myths about technical guardrails:

- **Myth 1: "Our vendor provides all the technical guardrails we need."**
Reality: Vendors provide features, not governance. You must configure, test, and monitor those features to align with your policies and risk tolerance.
- **Myth 2: "Technical guardrails make the system 100 percent safe."**
Reality: No guardrail is perfect. They reduce risk but do not eliminate it. Defense in depth—multiple layers of guardrails—is essential.
- **Myth 3: "Setting up guardrails once is sufficient."**
Reality: Threats evolve, models drift, and risks change. Guardrails require continuous monitoring, testing, and updating.
- **Myth 4: "Technical guardrails are too complex for non-technical managers to understand."**
Reality: You do not need to code them, but you must understand what they do, why they matter, and how to measure their effectiveness. That is your role.

Use this reality check to set realistic expectations and maintain accountability. Technical guardrails are powerful tools, but they are not magic. They require investment, expertise, and ongoing attention.

5.15 Summary: From Governance to Code

This chapter has translated governance principles into technical reality. You have learned how dataset validation ensures quality and integrity at the source, how prompt filtering and input validation prevent adversarial manipulation, how output screening and content moderation provide a final safety check, how red-team simulation exposes vulnerabilities, how automated bias detection enables continuous fairness testing, how interpretability tools make model behavior transparent, how rate limiting and abuse prevention protect system availability, how data isolation enforces multi-tenant security, how monitoring and logging provide visibility, and how secure model deployment protects your intellectual property.

The manager's role is not to implement these guardrails yourself, but to ensure they exist, are functioning, and are measured. Use the playbook of questions in this chapter to hold your technical teams accountable. Establish KPIs for each guardrail category and regularly review them. Treat technical guardrails as infrastructure, not optional add-ons. When governance policies and technical controls are aligned, you create a system that is both innovative and trustworthy—a system that leadership can defend, regulators can audit, and users can trust.

6 Human-in-the-Loop: The Ultimate Safety Net

6.1 When Automation Goes Too Far

Your healthcare organization deployed an AI triage system to prioritize patient calls based on symptom severity. The system was fast, consistent, and reduced wait times. Then a patient called, reporting chest pain and shortness of breath. The AI classified it as "moderate priority" because the patient was young and had no documented heart disease. The call sat in the queue for 45 minutes. By the time a nurse reviewed it, the patient had called 911 and was en route to the emergency room with a heart attack. The AI made a statistically reasonable decision based on population data, but it missed critical context that a human would have caught immediately. The incident review was clear: full automation without human oversight is not safe for life-critical decisions. This is why human-in-the-loop is not optional—it is the ultimate safety net.

Human-in-the-loop (HITL) means that human judgment is integrated into AI workflows at critical decision points. Humans do not simply observe the system; they actively review, validate, override, and improve it. HITL is essential for high-stakes use cases where errors have serious consequences, where context matters more than patterns, and where accountability demands that a human can explain and defend every decision. This chapter shows you how to design HITL mechanisms that protect safety and ethics without creating bottlenecks. You will learn about escalation thresholds, exception-review patterns, annotation feedback loops, and how to calibrate human oversight as systems scale. The goal is not to slow down AI, but to keep it aligned with mission and values even under pressure.

6.2 Why Human Judgment Remains Central

AI systems excel at pattern recognition, speed, and consistency. They can process thousands of cases in seconds, apply rules uniformly, and operate around the clock. But they lack human judgment: the ability to recognize when statistical patterns do not apply, to weigh competing values, to understand context and intent, and to accept moral and ethical responsibility. In high-stakes domains—healthcare, criminal justice, financial decisions, safety-critical systems—human judgment is not optional.

There are three reasons why human oversight remains essential. First, **AI systems are probabilistic, not deterministic**. They make predictions based on likelihood, not certainty. A prediction that is 95 percent confident is still wrong 5 percent of the time. Humans are needed to recognize when the 5 percent edge case is occurring and to intervene before harm occurs. Second, **AI systems lack common sense and contextual understanding**. They can miss signals that are obvious to humans: a patient downplaying symptom due to fear, a loan applicant with unusual but legitimate circumstances, or a safety sensor reading that is technically within range but contextually alarming.

Third, **accountability requires human ownership**. When an AI system makes a harmful decision, someone must be accountable. That cannot be the model. It must be a person who reviewed the decision, exercised judgment, and took responsibility. Regulators, courts, and the public will not accept "the algorithm decided" as an answer. Human-in-the-loop is how you maintain that chain of accountability. It ensures that every high-stakes decision has a human signature, a human review trail, and a human who can explain the rationale.

6.3 Three Models of Human-in-the-Loop Integration

HITL is not a single pattern. There are multiple ways to integrate human judgment into AI workflows, each with different trade-offs between safety, speed, and cost. The three primary models are **human-in-command, human-on-the-loop, and human-in-the-loop**.

Human-in-command means the AI system provides recommendations, but a human makes every final decision. The AI is advisory only. This is the highest level of human oversight and is appropriate for the most critical use cases: clinical diagnosis support, sentencing recommendations, high-value financial approvals. The advantage is maximum safety and accountability. The disadvantage is that it does not scale: every case requires human review, which is time-consuming and expensive. Use this model when errors are unacceptable and when the stakes justify the cost.

Human-on-the-loop means the AI system makes decisions autonomously, but humans monitor the system and can intervene when necessary. This is common in semi-autonomous systems like manufacturing robots, self-driving vehicles in controlled environments, or fraud detection systems. Humans do not review every decision, but they watch for anomalies, exceptions, or degradation. They have the authority to pause the system, override decisions, or escalate issues. This model balances automation and oversight, scaling better than human-in-command while maintaining a safety valve.

Human-in-the-loop (narrow definition) means the AI system escalates certain decisions to humans based on predefined criteria: low confidence, high stakes, edge cases, or user requests. Routine, high-confidence decisions are fully automated. Ambiguous or risky decisions are routed to human reviewers. This is the most scalable model for high-volume use cases: customer support routing, document processing, content moderation. It automates the majority of cases while ensuring that difficult or risky cases get human attention. The key is defining clear escalation thresholds so the right cases are reviewed without overwhelming human reviewers.

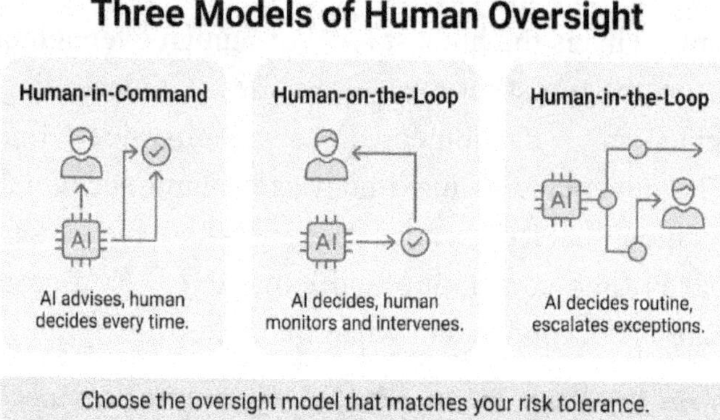

6.4 Designing Escalation Thresholds

The effectiveness of human-in-the-loop depends on escalation thresholds: the rules that determine which decisions are automated and which are routed to human review. Well-designed thresholds route the right cases without overwhelming reviewers. Poorly designed thresholds either miss risky cases (under-escalation) or drown reviewers in routine work (over-escalation).

Confidence-based thresholds are the simplest. If the model's confidence score falls below a threshold (e.g., 80 percent), escalate to human review. This works well when confidence scores are calibrated and reliable. However, confidence alone is insufficient. A model can be confidently wrong. Combine confidence thresholds with other signals.

Stake-based thresholds escalate decisions based on impact. For example, in a loan approval system, automatically approve loans under 10,000 dollars with high confidence, but escalate all loans over 50,000 dollars regardless of confidence. In a content moderation system, automatically remove clearly violating content, but escalate borderline cases. Stake-based thresholds ensure that high-impact decisions always get human review.

Novelty-based thresholds detect when the model encounters situations that are significantly different from its training data. If the input features fall outside the distribution the model was trained on, escalate. This prevents the model from confidently extrapolating into unfamiliar territory. Novelty detection requires instrumentation to measure feature distributions and flag outliers.

User-requested escalation allows users to request human review. This is especially important for systems that affect individuals: loan denials, job application rejections, medical triage. Give users a clear, easy path to say, "I want a human to review this." This provides both a safety valve and valuable feedback: repeated escalation requests on certain types of cases may indicate a model deficiency.

Combine multiple threshold types to create layered escalation logic. For example: escalate if confidence is below 80 percent OR if the decision stakes are above a threshold OR if the case is a novelty OR if the user requests review. Test and calibrate thresholds regularly. If 50 percent of cases are escalating, thresholds may be too aggressive. If serious errors are slipping through, thresholds may be too lenient.

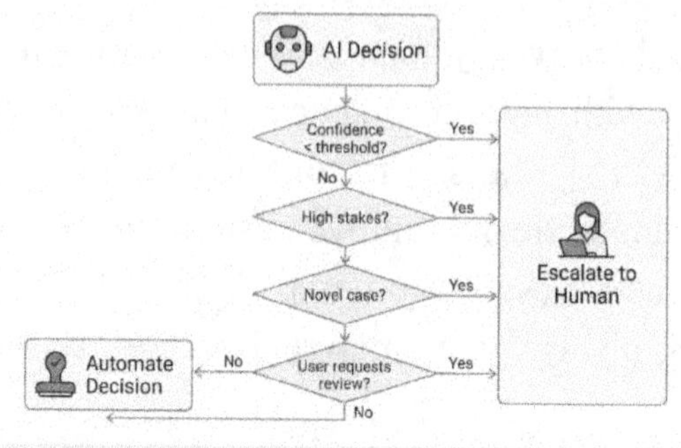

Escalation Threshold Framework

Use multiple signals to route decisions correctly.

6.5 Patterns: Making Human Review Efficient

When cases are escalated to human reviewers, the review process must be efficient, consistent, and well-supported. Exception review patterns are workflows and interfaces designed to help human reviewers make fast, accurate decisions without drowning in complexity.

Structured review interfaces present reviewers with all necessary information in a clear, standardized format. Include: the AI's recommendation and confidence score, the key factors that influenced the recommendation (feature importances or explanations), relevant historical context (previous interactions, prior decisions), policy guidance (which rules or criteria apply), and clear action options (approve, deny, modify, escalate further). Avoid overwhelming reviewers with raw data. Present curated, decision-relevant information.

Decision support tools help reviewers make consistent choices. Provide reference materials: policy documents, decision trees, examples of similar cases and how they were resolved. Offer calculators or lookup tools that reviewers can use to verify information or apply complex rules. For example, a loan reviewer might have access to a debt-to-income calculator or a creditworthiness reference table.

Batch review workflows group similar cases together so reviewers can process them efficiently. If 20 cases were flagged for the same reason (e.g., low confidence on a specific feature), present them together. This allows reviewers to develop intuition and pattern recognition. It also makes it easier to identify systemic issues: if all 20 cases suggest the model is struggling with a particular scenario, that is a signal to retrain or update the model.

Time-boxing and prioritization ensure that urgent cases are reviewed first. Assign priority levels to escalated cases (critical, high, medium, low) based on urgency and impact. Route critical cases to senior reviewers or subject matter experts. Set time targets: critical cases reviewed within 1 hour, high-priority within 4 hours, medium within 24 hours. Monitor review queue metrics: average wait time, backlog size, and reviewer workload. If queues grow too long, either add reviewer capacity or tighten escalation thresholds to reduce volume.

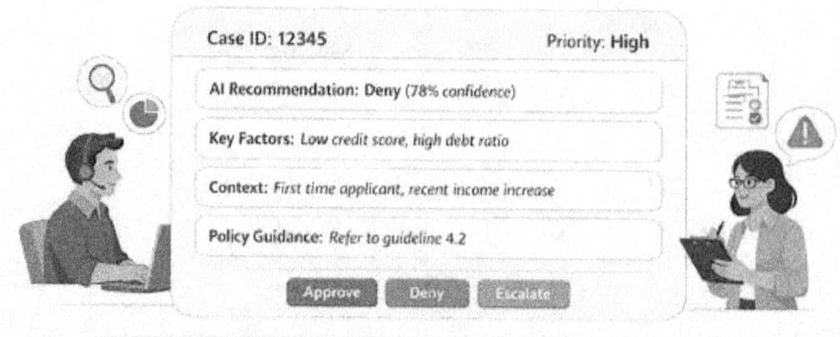

Give reviewers everything they need in a clear, standardized format.

6.6 Improving Models Through Human Review

Human review is not just about catching errors; it is also an opportunity to improve the AI system. Annotation and feedback loops capture human decisions and explanations, feeding them back into the training process to make models smarter and better aligned over time.

Annotation means that when a human reviewer makes a decision, they also provide structured feedback about why. For example, if a reviewer overrides the AI's recommendation to deny a loan and approves it instead, they select a reason: "AI underweighted recent income increase," or "Applicant provided additional context not in training data." These annotations create labeled data that can be used to retrain or fine-tune the model. Over time, the model learns from human corrections.

Feedback loop architecture connects human review systems to model training pipelines. When reviewers make decisions, those decisions are logged along with annotations. Periodically (e.g., monthly or quarterly), data scientists analyze the feedback: Which types of cases are most frequently overridden? What patterns emerge? Are there systematic biases in the model's recommendations? Use this analysis to update training data, adjust feature engineering, or retrain models. The human review process becomes a continuous improvement engine.

Disagreement analysis is especially valuable. When multiple reviewers see the same case, do they agree? If not, why? Disagreement can reveal ambiguous policies, insufficient guidance, or cases where human judgment itself is inconsistent. Use disagreement as a signal to clarify policies, provide additional training to reviewers, or identify cases where the model should not be making decisions at all.

Active learning takes feedback loops further by having the AI system intelligently select which cases to send for human annotation. Instead of randomly sampling, the model identifies cases where it is most uncertain or where human feedback would be most valuable. This accelerates learning: the model improves faster because it is learning from the most informative examples, not just the most common ones.

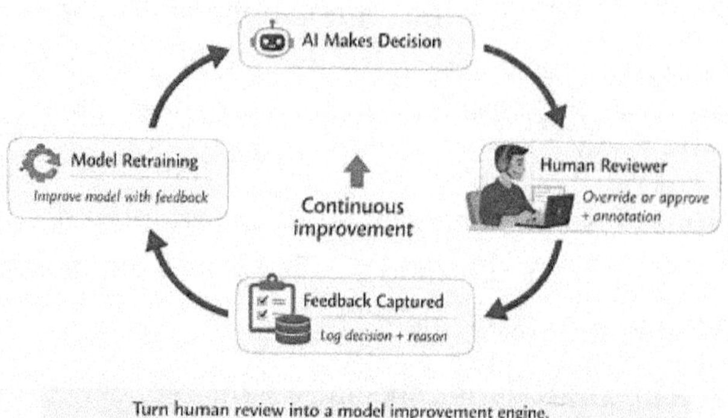

6.7 Case : Human-in-the-Loop in Healthcare Triage

To make these concepts concrete, consider a real-world application: AI-powered triage in a hospital emergency department. The system analyzes patient symptoms, vitals, and medical history to assign a priority level (critical, urgent, standard, low). Here is how human-in-the-loop is implemented.

Model selection: The organization uses a human-on-the-loop model. The AI assigns initial priority, but nurses monitor all assignments and can override or escalate.

Escalation thresholds: The system automatically escalates to immediate nurse review if: the patient reports chest pain, difficulty breathing, severe bleeding, loss of consciousness, or any symptom keywords associated with life-threatening conditions; if the patient's vitals are outside safe ranges (heart rate, blood pressure, oxygen saturation); if the AI's confidence is below 85 percent; or if the patient or family requests immediate attention.

Review interface: When a case is escalated, the nurse sees a summary screen showing: patient demographics and chief complaint, AI-assigned priority and confidence score, key symptom keywords and vital signs, relevant medical history (chronic conditions, recent visits), and policy guidance (triage protocols for specific symptoms). The nurse can approve the AI's priority, adjust it up or down, or escalate to a physician.

Annotation: When a nurse overrides the AI, they select a reason from a dropdown: "AI missed critical symptom," "Patient context requires higher priority," "Vitals more concerning than AI assessed," or "Other (free text)." These annotations are logged.

Feedback loop: Every month, the clinical informatics team reviews override patterns. They discover that the AI consistently prioritizes elderly patients with vague symptoms (fatigue, weakness) who turn out to have serious conditions. The team adjusts the model to increase sensitivity for elderly patients with non-specific symptoms and adds a feature for age-adjusted risk scoring. Over six months, overrides for this scenario drop by 40 percent, indicating the model has learned from human feedback.

Outcome: The system processes 80 percent of triage cases automatically, freeing nurses to focus on complex or ambiguous cases. Critical cases are escalated immediately to ensure no delays. The feedback loop continuously improves the model, reducing overrides and increasing trust. Human-in-the-loop does not slow down the system; it makes it safer and smarter.

6.8 Balancing Automation and Oversight Bottlenecks

A common fear among humans in the loop is that it will create bottlenecks, slowing operations and negating the benefits of automation. This happens when escalation thresholds are miscalibrated, review processes are inefficient, or human capacity is insufficient. The key is to design HITL systems that balance automation and oversight without sacrificing either speed or safety.

Start with data-driven threshold calibration. Before deploying HITL, analyze historical data: How many cases would escalate under different threshold settings? What is the distribution of confidence scores? What percentage of cases are high-stakes? Use this analysis to set initial thresholds that route a manageable volume to human review. For example, if you have 10,000 cases per day and 5 human reviewers, each capable of reviewing 50 cases per day, you can handle 250 escalations per day—2.5 percent of total volume—set thresholds to target that escalation rate.

Monitor and adjust dynamically. After deployment, track escalation volume, review queue wait times, and reviewer workload. If queues are growing, tighten thresholds (require higher confidence for escalation) or add reviewer capacity. If serious errors are slipping through, loosen thresholds. Treat threshold calibration as an ongoing operational task, not a one-time design decision.

Automate the easy decisions completely. Use the most permissive automation for low-stakes, high-confidence cases. For example, in customer support, if the AI is 95 percent confident and the request is routine (a password reset or tracking inquiry), automate it fully. Reserve human review for high-stakes (billing disputes, complaints) or low-confidence cases. This maximizes the benefits of automation while protecting safety.

Empower reviewers with efficiency tools. As discussed in Section 5.4, structured interfaces, batch workflows, and decision support tools accelerate human review. If reviewers spend 10 minutes per case instead of 2 minutes, your capacity drops by 80 percent. Invest in user experience for reviewers, not just end users.

Finally, **accept that some use cases cannot scale with full HITL**. If every decision requires human review and you have millions of decisions per day, having humans-in-the-loop may not be viable. In such cases, either accept higher risk with automation and technical guardrails, or reconsider whether the use case is appropriate for AI at all. Not every problem should be solved with automation.

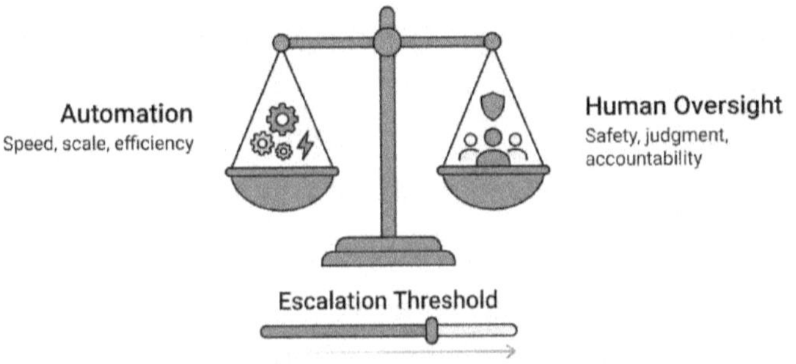

Calibrate thresholds to balance speed and safety without bottlenecks

6.9 Training and Empowering Human Reviewers

Human-in-the-loop is only as good as the humans in the loop. Reviewers need training, tools, support, and authority to do their jobs effectively. Organizations often underinvest in reviewer training, assuming that domain expertise is sufficient. In reality, reviewing AI decisions requires a different skill set than performing the task manually.

AI literacy training helps reviewers understand what the AI is doing and its limitations. Reviewers should know: what data the model was trained on, what features it uses, what types of errors it is prone to, and how to interpret confidence scores and explanations. This prevents over-reliance on AI recommendations (automation bias) and under-reliance (dismissing AI input without consideration). Training should be practical: use real examples from your system to show both correct and incorrect AI decisions.

Policy and procedure training ensures that reviewers apply consistent standards. When should they approve the AI's recommendation? When should they override? When should they escalate further? Document clear criteria and provide examples. Role-play scenarios where reviewers must make judgment calls and discuss the reasoning. Update training materials regularly as policies evolve and as feedback loops reveal new patterns.

Tool and interface training ensure reviewers can use the review system efficiently. Walk them through the interface, show them how to access reference materials, and demonstrate workflows for common scenarios. Provide quick-reference guides and keyboard shortcuts. The easier the tools are to use, the faster and more accurately reviewers can work.

Psychological and ethical support is important, especially for reviewers in high-stress domains (content moderation, medical triage, criminal justice). Reviewing difficult cases repeatedly can lead to burnout, compassion fatigue, or moral distress. Provide access to mental health support, rotate reviewers through different tasks, and create a culture where it is acceptable to take breaks or escalate cases that are too difficult. Recognize and reward reviewers for their contributions: they are not just "checking the AI's work"; they are the final line of defense against harm.

Finally, **empower reviewers with authority**. If reviewers believe their decisions will be second-guessed or ignored, they will disengage. Make it clear that reviewers have the authority to override AI decisions and that their judgment is trusted. When reviewers flag systemic issues (e.g., "The AI keeps making this same mistake"), ensure those issues are investigated and addressed. Reviewers are your early warning system. Listen to them.

Reviewer Training Framework

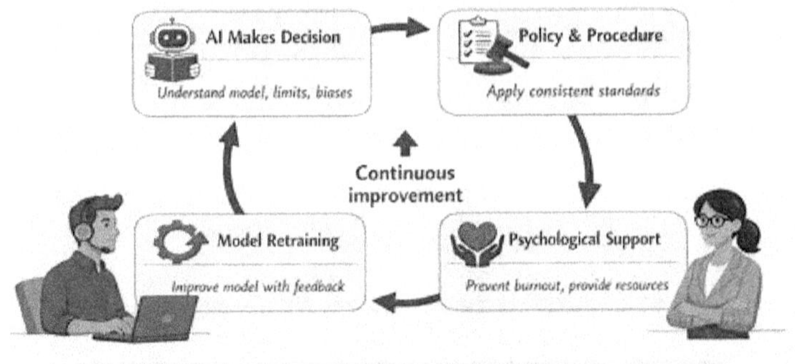

6.10   Case: Human-in-the-Loop in Benefits Eligibility

Consider another case: a government agency uses AI to determine eligibility for social benefits (unemployment, disability, housing assistance). The stakes are high: incorrect denials can cause significant harm to vulnerable individuals. Here is how HITL is implemented.

Model selection: Human-in-the-loop (narrow definition). Straightforward cases that clearly meet or clearly fail eligibility criteria are automated. Edge cases, borderline situations, and applicants who request review are escalated to case workers.

Escalation thresholds: Escalate if: AI confidence is below 90 percent; applicant income or circumstances are borderline (within 10 percent of eligibility threshold); applicant has unusual circumstances not well-represented in training data (recent job loss, medical emergency, family situation); or applicant requests human review.

Review interface: Case workers see: applicant demographics and application details, AI recommendation (approve/deny) and confidence score, key eligibility factors (income, family size, employment status), explanation of which factors drove the AI's decision, and policy references (relevant sections of eligibility guidelines). Case workers can approve, deny, request additional documentation, or schedule an interview.

Annotation: When case workers override the AI, they document reasons: "AI did not account for medical expenses," "Applicant provided proof of extenuating circumstances," "Policy exception applies." Annotations are structured and include references to policy sections.

Feedback loop: Quarterly, the agency's data team reviews overrides and disagreements. They identify policy areas where the AI struggles (e.g., handling temporary disability vs. permanent disability) and update the model or adjust business rules accordingly. They also identify cases where policy guidance is ambiguous and work with policy teams to clarify.

Outcome: The system processes 70 percent of applications automatically, providing immediate decisions for clear-cut cases. This reduces wait times from weeks to hours for most applicants. Case workers focus on complex cases where human judgment is essential. Feedback loops ensure the model improves and stays aligned with policy changes. Applicants can request human review to ensure transparency and fairness. The result is faster service, more consistent decisions, and protection for vulnerable populations.

6.11 Handling Disagreements Between AI and Humans

When humans and AI disagree, who is right? This question is central to the design of HITL. Sometimes the AI is correct, and the human is wrong (humans make mistakes, have biases, or lack information). Sometimes the human is correct, and the AI is wrong (the AI missed context, made a statistical error, or encountered a novel situation). Sometimes both are partially right, and the truth lies in between.

Document all disagreements. When a human overrides the AI, log the case, the AI's recommendation, the human's decision, and the reason for the override. Track override rates: What percentage of escalated cases are overridden? Is the rate increasing or decreasing over time? High override rates may indicate that the AI is not performing well or that escalation thresholds are set incorrectly.

Investigate patterns in disagreements. Are overrides clustered around certain types of cases, certain reviewers, or certain times? Clustering by case type suggests a model deficiency. Clustering by reviewer suggests inconsistent training or bias. Clustering by time suggests the presence of external factors (policy changes, data drift, seasonal patterns).

Establish ground truth where possible. For some use cases, you can determine who was right after the fact. For example, in fraud detection, did the transaction turn out to be fraudulent? In medical triage, did the patient's condition require urgent care? Use ground truth to score both AI and human decisions. This provides objective feedback and reveals whether AI or human judgment is more accurate for different case types.

Use disagreements to improve the system. When the AI is consistently wrong in a given scenario, retrain the model or adjust the features. When humans are frequently wrong, provide additional training or update policies. When both are struggling, the scenario may be inherently ambiguous, and you may need to adjust the process (e.g., require multi-reviewer consensus, escalate to senior experts, or accept higher uncertainty).

Finally, **create a culture of learning, not blame**. Disagreements are data, not failures. Treat them as opportunities to improve both the AI and the human processes. Avoid framing overrides as "the AI was bad" or "the human was wrong." Instead, ask: "What can we learn from this disagreement to make better decisions in the future?"

Treat disagreements as learning opportunities, not failures.

6.12 Scaling Human-in-the-Loop Maintain Quality

As AI systems grow and handle more volume, humans-in-the-loop must scale too. The challenge is to maintain review quality while handling increased volume. There are several strategies for scaling HITL without degradation.

Tiered reviewer models use different levels of expertise for different case types. Routine escalations go to junior reviewers or generalists. Complex or high-stakes cases go to senior reviewers or subject-matter experts. This maximizes efficiency: you do not need expensive experts to review every case, but they are available when needed. Implement clear routing rules based on case complexity, stakes, and novelty.

Distributed review networks expand capacity by using a larger, geographically distributed pool of reviewers. This can include remote workers, contractors, or crowdsourcing platforms (for appropriate use cases). Distributed networks provide flexibility and scalability, but require careful quality control: training programs, performance monitoring, spot checks, and consensus mechanisms (multiple reviewers per case for critical decisions).

Automated quality assurance monitors reviewer performance to catch errors or inconsistencies. Randomly sample reviewed cases and have senior reviewers review them. Measure inter-reviewer agreement: do different reviewers make consistent decisions on similar cases? Flag reviewers with low agreement or high error rates for additional training or supervision. Use these metrics not punitively, but as indicators of where support is needed.

Hybrid workflows combine different oversight models for different parts of the process. For example, use human-in-the-loop for initial decisions but human-on-the-loop for ongoing monitoring. Or automate initial screening, but require a human-in-command for final approvals. Hybrid workflows allow you to allocate human effort where it matters most.

Continual threshold optimization adjusts escalation rules as the model improves. If feedback loops and retraining reduce error rates and increase confidence, you can gradually tighten escalation thresholds, reducing the volume of cases requiring human review. Monitor outcomes carefully: if tightening thresholds leads to increased errors slipping through, loosen them again. Treat threshold optimization as a continuous process, not a one-time setting.

6.13   Playbook: Implementing Human-in-the-Loop

You have seen the principles, models, and strategies. Now, how do you actually implement HITL in your organization? Here is a manager-ready playbook.

Step 1: Assess use case risk and volume. For each AI use case, determine: What is the potential harm if the AI makes an error? What is the decision volume (cases per day/month)? What level of human oversight is appropriate (human-in-command, human-on-the-loop, human-in-the-loop)? Document this assessment and share it with your governance team.

Step 2: Design escalation logic. Define thresholds: confidence levels, stake levels, novelty detection, user-requested review. Start conservatively (escalate more cases) and tighten over time as you gain confidence. Document the logic clearly so technical teams can implement it.

Step 3: Build or procure review tools. Design structured review interfaces that present all necessary information clearly. Ensure tools support batch processing, prioritization, and annotation. If building custom tools is not feasible, evaluate third-party platforms that provide case management and review workflows.

Step 4: Recruit and train reviewers. Identify who will perform human review: existing staff, new hires, contractors? Develop training programs covering AI literacy, policy, tools, and support. Pilot the training with a small group and iterate based on feedback.

Step 5: Implement feedback loops. Ensure that human decisions and annotations are logged and accessible to your data science team. Establish a cadence (monthly or quarterly) for analyzing feedback and retraining models. Assign ownership: who is responsible for closing the loop?

Step 6: Pilot the HITL system. Deploy to a limited scope (one team, one region, one use case subset). Monitor escalation volume, review queue times, override rates, and user feedback. Identify bottlenecks and pain points. Iterate on thresholds, tools, and processes.

Step 7: Scale and optimize. Once the pilot is successful, roll out to full scope. Continue monitoring and optimizing thresholds, reviewing capacity, and model performance. Treat HITL as a living system that evolves with your AI program.

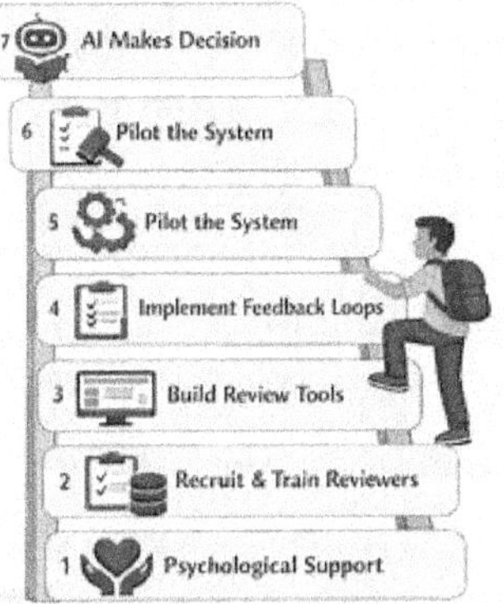

Follow this sequence to integrate human judgment into AI workflows.

6.14 Pitfalls in Human-in-the-Loop Implementation

Even well-designed HITL systems can fail. Here are five common pitfalls and how to avoid them.

- **Pitfall 1: Automation bias**. Reviewers over-rely on AI recommendations and rubber-stamp decisions without true review. **Avoidance:** Train reviewers on AI limitations. Present cases in ways that require active engagement (e.g., hide the AI recommendation until the reviewer makes their own assessment). Monitor override rates: if they are too low, investigate whether reviewers are genuinely reviewing or just approving.

- **Pitfall 2: Alert fatigue**. Too many escalations overwhelm reviewers, leading to rushed decisions or disengagement. **Avoidance:** Calibrate thresholds to keep escalation volume manageable. Prioritize cases so reviewers focus on the most critical first. Provide adequate reviewer capacity.

- **Pitfall 3: Lack of feedback loop**. Human decisions are logged but never analyzed or fed back into model training. **Avoidance:** Assign explicit ownership for closing the feedback loop. Schedule regular reviews of human override data. Ensure data scientists have access to annotated feedback and use it in retraining.

- **Pitfall 4: Inconsistent review quality**. Different reviewers make different decisions on similar cases. **Avoidance:** Provide clear, documented policies and decision criteria. Measure inter-reviewer agreement. Offer calibration sessions where reviewers discuss difficult cases and align on standards.

- **Pitfall 5: Reviewer burnout**. Reviewing difficult, high-stakes cases repeatedly is emotionally taxing. **Avoidance:** Rotate reviewers through different tasks. Provide mental health and peer support. Recognize and value reviewers' contributions. Monitor for signs of burnout and intervene early.

6.15 Human Judgment as the Ultimate Guardrail

This chapter has shown you why human judgment remains central to responsible AI, even as automation scales. You have learned the three models of human oversight (human-in-command, human-on-the-loop, human-in-the-loop), how to design escalation thresholds that route the right cases to human review, how to build efficient exception review workflows, how to create annotation feedback loops that improve models over time, and how to scale human oversight without sacrificing quality or creating bottlenecks.

Human-in-the-loop is not about slowing down AI. It is about keeping AI aligned with mission, ethics, and reality. It ensures that statistical patterns do not override common sense, that context matters, and that accountability remains with people, not algorithms. The organizations that succeed with AI at scale will be those that master the integration of human and machine intelligence—using automation for efficiency and human judgment for safety. As an IT manager, your role is to design, implement, and continuously optimize HITL systems to make them a seamless, effective part of your AI operating model. When done right, human-in-the-loop is not a bottleneck. It is your ultimate safety net.

7 Risk Assessment and Triage Models

Opening Scenario: When All AI Gets Treated the Same

Your organization has 23 AI use cases in production, ranging from an internal chatbot that answers HR questions to a customer credit scoring system that determines loan approvals. Your governance team insists that every use case go through the same review process: full fairness assessment, red team testing, legal review, and executive sign-off. The internal chatbot team is frustrated—they are stuck in a three-month approval queue for a tool that only employees use and provides advisory information, not decisions. Meanwhile, the credit scoring team breezes through the same process because they submitted their paperwork first, even though their system makes automated decisions affecting thousands of customers. This is the problem with one-size-fits-all governance:

it either over-controls low-risk use cases or under-controls high-risk ones. Neither outcome is acceptable.

Risk assessment and triage are about making intentional, proportional decisions about where to invest your limited governance resources. Not all AI use cases carry the same risk. An internal knowledge assistant and a medical diagnosis tool should not be governed the same way. This chapter introduces a structured model for assessing AI risk by impact and likelihood, classifying use cases by criticality, and applying corresponding levels of control. You will learn the AI Risk Triage Framework—a practical tool that enables your organization to focus on guardrails where the stakes are highest while accepting managed risk elsewhere. The goal is not to eliminate all risk, but to understand it, prioritize it, and address it proportionally.

7.1 The Case for Risk-Based Governance

Risk-based governance means that the intensity of oversight, controls, and testing is calibrated to the level of risk a use case presents. High-risk use cases get rigorous review, extensive testing, mandatory human oversight, and frequent audits. Low-risk use cases get lightweight approval, basic checks, and periodic spot reviews. This approach is both practical and strategic. It is practical because you do not have infinite resources to govern every AI system with maximum rigor. It is strategic because it signals to leadership and regulators that you understand risk and are managing it thoughtfully.

The alternative to risk-based governance is either uniform high governance (which creates bottlenecks and slows innovation) or uniform low governance (which exposes the organization to unacceptable harm). Both are untenable at scale. Risk-based governance threads the needle: it enables speed where speed is safe and imposes discipline where discipline is essential. It also creates a shared language for discussing trade-offs. Instead of abstract debates about whether a use case is "too risky," you can point to specific risk dimensions, score them objectively, and apply predefined rules.

For IT managers, risk-based governance is a familiar concept in other domains, such as cybersecurity, operational risk management, and project management. The principles are the same: assess the likelihood and impact of adverse events, prioritize accordingly, and allocate resources. The challenge in AI is that risk dimensions are broader than traditional IT risk. You must account not only for technical failures but also for fairness, transparency, accountability, and societal impact. This chapter gives you a structured way to do that.

7.2 Key Risk Dimensions in AI Systems

To assess risk, you need a clear understanding of the dimensions that matter. For AI systems, there are six key risk dimensions: **impact on individuals, impact on organization, automation level, data sensitivity, model complexity, and regulatory exposure**.

Impact on individuals measures how AI decisions affect people's lives. Does the system influence hiring, lending, healthcare, criminal justice, or access to essential services? If the system makes an error, does it harm someone's livelihood, safety, dignity, or legal rights? High individual impact means high risk. An AI chatbot that recommends movies has low individual impact. An AI system that denies disability benefits has a high individual impact.

Impact on organization measures potential harm to your organization: financial loss, reputational damage, legal liability, and operational disruption. A system that processes millions of dollars in transactions has a high organizational impact. A system that generates internal reports has low organizational impact. Consider both direct costs (fines, settlements) and indirect costs (brand damage, lost customers, regulatory scrutiny).

Automation level measures the degree of human oversight. Is the system advisory-only (provides recommendations for humans to review)? Does it make decisions that humans can override? Does it take automated actions with no human in the loop? Higher automation means higher risk because errors propagate without human intervention. A fraud detection system that flags transactions for review is lower risk than one that automatically freezes accounts.

Data sensitivity measures the nature of data that the system uses or exposes. Does it involve personally identifiable information (PII), protected health information (PHI), financial data, children's data, or data from vulnerable populations? Does it cross borders or jurisdictions with strict data regulations? Higher data sensitivity means higher privacy and compliance risk. A system trained on public data has lower sensitivity than one using customer health records.

Model complexity measures how difficult it is to understand, test, and validate the system. Simple rule-based systems or interpretable models (decision trees, linear models) are easier to audit and explain. Complex deep learning models are harder to interpret and test. Higher complexity does not automatically mean higher risk, but it does mean higher uncertainty and more effort required to verify behavior. Complexity becomes a risk multiplier:

when combined with high impact or high automation, it increases overall risk.

Regulatory exposure measures the extent to which the use case is subject to existing or emerging regulations. Is it in a highly regulated sector (healthcare, finance, government)? Does it involve protected characteristics (race, gender, disability) that trigger anti-discrimination laws? Are there specific AI regulations that apply (e.g., the EU AI Act, sector-specific rules)? Higher regulatory exposure means higher compliance risk and the potential for enforcement actions, audits, or legal challenges.

Six Risk Dimensions for AI Systems

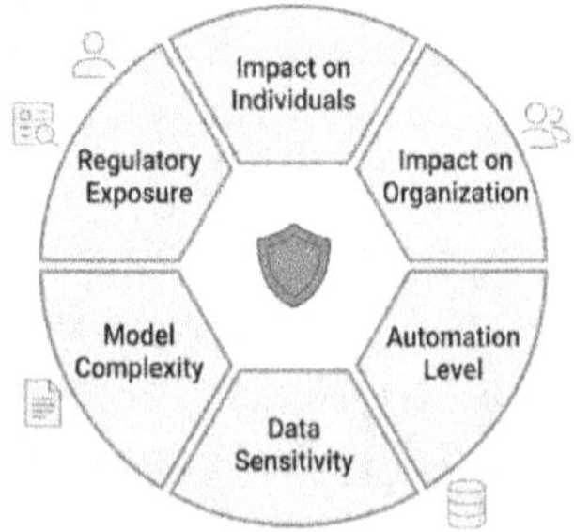

Assess all six dimensions to understand overall AI risk.

7.3 The AI Risk Triage Framework

The AI Risk Triage Framework is a practical tool for classifying use cases into risk tiers and assigning appropriate governance. It uses a simple scoring system based on the six risk dimensions, maps scores to three risk tiers (low, medium, high), and defines what governance is required for each tier.

Step 1: Score each risk dimension. For each of the six dimensions, assign a score of 1 (low risk), 2 (medium risk), or 3 (high risk) based on defined criteria. For example:

- **Impact on individuals:** 1 = No impact on people's lives (internal tool, informational only); 2 = Moderate impact (influences decisions but not binding, easily reversible); 3 = High impact (affects access to services, employment, safety, legal rights).

- **Impact on organization:** 1 = Minimal financial or reputational impact; 2 = Moderate impact (up to $1M exposure, localized reputational risk); 3 = High impact (>$1M exposure, enterprise-wide reputational risk, regulatory enforcement).

- **Automation level:** 1 = Advisory only (human makes all final decisions); 2 = Semi-automated (AI decides, human can override); 3 = Fully automated (AI decides and acts without human review).

- **Data sensitivity:** 1 = Public or non-sensitive data; 2 = Internal business data, limited PII; 3 = Highly sensitive (PHI, financial, children's data, protected characteristics).

- **Model complexity:** 1 = Simple, interpretable models; 2 = Moderate complexity (standard ML models, some explainability); 3 = High complexity (deep learning, limited interpretability).

- **Regulatory exposure:** 1 = Low regulation, minimal legal risk; 2 = Moderate regulation (sector-specific rules apply); 3 = High regulation (strict AI laws, protected characteristics, cross-border).

Step 2: Calculate total risk score. Sum the scores across all six dimensions. The total score ranges from 6 (lowest risk) to 18 (highest risk).

Step 3: Map to risk tier. Use thresholds to assign a tier:

- **Low risk:** Total score 6–9
- **Medium risk:** Total score 10–14
- **High risk:** Total score 15–18

Step 4: Apply corresponding governance. Each tier has predefined governance requirements (detailed in the next section).

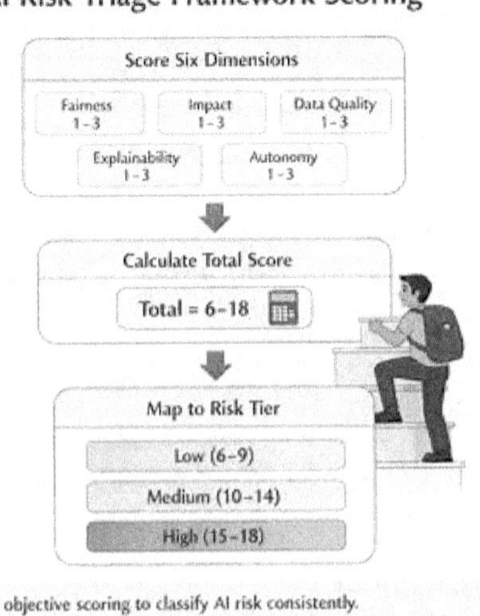

7.4 Governance Requirements by Risk Tier

Once you have classified a use case into a risk tier, apply the corresponding governance requirements. These requirements define what approvals, testing, controls, and monitoring are mandatory for that tier.

Low-risk tier (score 6–9):

- **Approval:** Business owner or product manager sign-off; no governance committee required.
- **Documentation:** Basic use case description, data sources, model type.
- **Testing:** Standard unit tests, basic quality checks.
- **Fairness review:** Not required (unless specific concerns are flagged).
- **Security review:** Standard security checklist.
- **Human oversight:** Optional; advisory mode acceptable.
- **Monitoring:** Periodic spot checks (quarterly or semi-annually).
- **Audit frequency:** Every two years or as needed.

Examples: Internal knowledge assistant, document summarization tool, meeting scheduler, simple recommendation engine for non-critical content.

Medium-risk tier (score 10–14):

- **Approval:** AI Working Group review and approval required.
- **Documentation:** Detailed use case, data lineage, model card, risk assessment.
- **Testing:** Comprehensive testing, including bias testing, edge cases, and red-team scenarios.
- **Fairness review:** Required if the system involves people or groups; document results.
- **Security review:** Full security assessment, penetration testing.

- **Human oversight:** Human-in-the-loop for exceptions or high-stakes decisions.
- **Monitoring:** Regular performance monitoring (monthly); automated alerts for anomalies.
- Audit frequency: Annually.

Examples: Customer support routing, fraud detection (with human review), internal HR tools, supply chain optimization, marketing personalization.

High-risk tier (score 15–18):

- **Approval:** AI Steering Committee and legal/compliance review required.
- **Documentation:** Comprehensive documentation, including ethics assessment, legal opinion, and risk mitigation plan.
- **Testing:** Rigorous testing, including fairness, explainability, adversarial robustness, and safety testing; a third-party audit may be required.
- **Fairness review:** Mandatory; ongoing fairness monitoring required.
- **Security review:** Comprehensive security and privacy assessment; penetration testing and red-team exercises.
- **Human oversight:** Human-in-command or human-in-the-loop with strict escalation thresholds; no fully automated decisions without review.
- **Monitoring:** Continuous monitoring with real-time alerts; performance dashboards reviewed weekly.
- **Audit frequency:** Semi-annually or more frequently if required by regulation.

Examples: Credit scoring and lending decisions, medical diagnosis support, criminal justice risk assessment, hiring and promotion tools, child welfare case management, and autonomous safety-critical systems.

Governance Requirements by Risk Tier

Requirement	Low Risk	Medium Risk	High Risk
Approval	Business owner	AI Working Group	Steering Committee
Documentation	Documentation	Documentation	Documentadion
Testing	Testing	Testing	Testing
Fairness Review	Fairness	Fairness Review	Fairiness Review
Security Review	Security Review	Security Review	Security Review
Human Oversight	Human	Human Oversight	Human
Monitoring	Monitoring	Monitoring	Monitoring
Audit Frequency	1 year	6 months	1 month

Match governance intensity to risk level.

7.5 Applying the Framework: Scoring Examples

To make the framework concrete, let's score three example use cases and see how they map to risk tiers.

Example 1: Internal Knowledge Assistant

- **Impact on individuals:** 1 (advisory only, no binding decisions)
- **Impact on organization:** 1 (low financial or reputational risk)
- **Automation level:** 1 (advisory, humans make all decisions)
- **Data sensitivity:** 2 (internal business documents, some PII)
- **Model complexity:** 2 (standard language model, some explainability)
- **Regulatory exposure:** 1 (low regulatory risk)

- Total score: 8 → Low risk
- **Governance:** Business owner approval, basic documentation, standard testing, periodic spot checks.

Example 2: Fraud Detection System

- **Impact on individuals:** 2 (flags transactions for review, can cause inconvenience)
- **Impact on organization:** 3 (high financial impact if fraud is missed or false positives harm customers)
- **Automation level:** 2 (flags for human review, not fully automated)
- **Data sensitivity:** 3 (financial data, PII)
- **Model complexity:** 2 (standard ML model, interpretable)
- **Regulatory exposure:** 2 (financial regulations apply)
- Total score: 14 → Medium risk
- **Governance:** AI Working Group approval, detailed documentation, bias and security testing, human-in-the-loop for high-stakes flags, monthly monitoring, and annual audit.

Example 3: Automated Loan Approval

- **Impact on individuals:** 3 (affects access to credit, significant financial impact)
- **Impact on organization:** 3 (high financial and reputational risk, regulatory penalties)
- **Automation level:** 3 (fully automated decisions up to certain limits)
- **Data sensitivity:** 3 (financial data, credit history, PII)
- **Model complexity:** 2 (standard credit model, somewhat interpretable)
- **Regulatory exposure:** 3 (fair lending laws, AI regulations, financial regulations)

- Total score: 17 → High risk
- **Governance:** Steering Committee approval, comprehensive documentation, rigorous testing including fairness and explainability, third-party audit, human-in-the-loop for borderline cases, continuous monitoring, semi-annual audits.

These examples show how the framework produces differentiated governance. The internal assistant gets lightweight oversight. The fraud system gets moderate scrutiny. The loan system gets maximum rigor. This is proportional risk management.

7.6 Handling Edge Cases and Disagreements

The framework provides structure, but judgment is still required. There will be edge cases where scoring is ambiguous or where stakeholders disagree about the appropriate tier. Here is how to handle them.

Ambiguous scores: If a dimension falls between two levels (e.g., "Is this a 2 or a 3?"), document the uncertainty and score conservatively. When in doubt, score higher. It is better to govern initially over and relax controls later than to under-govern and discover a gap after an incident.

Borderline total scores: If a use case scores exactly at a tier boundary (e.g., score of 10, right at the low/medium boundary), consider context. Are there specific risks that justify moving up a tier? Are there mitigating controls that justify staying in the lower tier? Document the rationale.

Stakeholder disagreements: Different stakeholders may assess risk differently. Legal may see high regulatory risk; the business may see low organizational impact. Resolve disagreements through structured discussion: present the scoring, hear arguments, and have a designated decision-maker (typically the AI Working Group chair) make the final call. Document the disagreement and the decision.

Risk evolution: Risk can change over time. A use case may start as low risk (pilot with 50 users) but become medium or high risk as it scales (deployed to 50,000 users). Re-score use cases periodically (at least annually) and when significant changes occur (new features, new data sources, expanded scope). If a use case moves up a tier, apply the new governance requirements.

Multiple use cases in one system: If a single AI system serves multiple use cases with different risk profiles, score each use case separately and apply the highest tier's governance to the overall system. Alternatively, architect the system to isolate use cases so they can be governed independently.

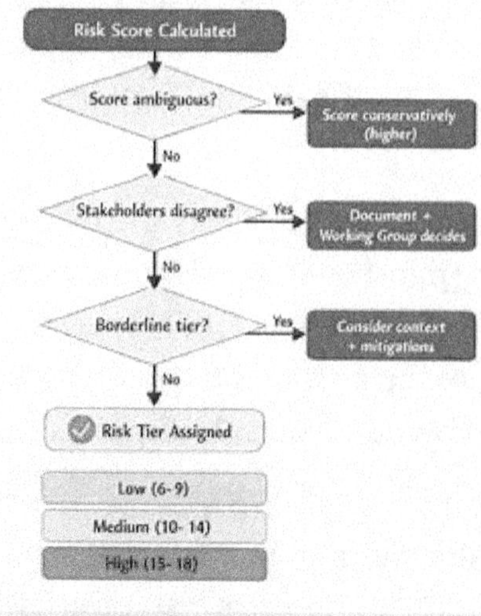

7.7 Risk Mitigation Strategies

Risk assessment identifies where risk exists. Risk mitigation reduces that risk to an acceptable level. For medium- and high-risk use cases, mitigation strategies should be documented as part of the approval process. Here are common mitigation strategies organized by risk dimension.

For high individual impact:

- Implement human-in-the-loop or human-in-command oversight.
- Provide explanations for decisions so individuals understand and can appeal.
- Create clear appeal or review processes for affected individuals.

- Conduct fairness testing and ongoing bias monitoring.

For high organizational impact:

- Implement phased rollout (pilot, then scale) to limit exposure.
- Establish financial caps (e.g., automate only decisions under a threshold).
- Require executive approval for full deployment.
- Maintain insurance or reserves for potential liabilities.

For a high automation level:

- Add human review for edge cases, low-confidence decisions, or high-stakes outcomes.
- Implement output screening and validation checks.
- Create circuit breakers that pause the system if anomalies are detected.
- Monitor closely and establish rapid rollback procedures.

For high data sensitivity:

- Use data minimization: collect and use only what is strictly necessary.
- Implement anonymization, pseudonymization, or differential privacy.
- Enforce strict access controls and audit all data access.
- Ensure compliance with data protection regulations (GDPR, HIPAA, etc.).

For high model complexity:

- Use interpretability tools to provide explanations.
- Conduct extensive testing, including adversarial and edge case testing.

- Consider using simpler, more interpretable models if acceptable performance can be achieved.
- Document model behavior thoroughly and maintain model cards.

For high regulatory exposure:

- Engage legal and compliance early in the design process.
- Conduct regulatory impact assessments.
- Build in controls and documentation that satisfy regulatory requirements.
- Monitor regulatory developments and update systems as rules evolve.

Document mitigation strategies for each medium- and high-risk use case. Verify that mitigations are implemented before deployment. Monitor their effectiveness after deployment.

Risk Mitigation Strategies by Dimension

Tailor mitigation strategies to specific risk dimensions.

7.8 Risk-Control Matrix: Aligning Risk & Resources

The Risk Triage Framework indicates the level of governance required. The Risk-Control Matrix helps you decide how to allocate resources across your AI portfolio. Not all use cases can or should be high priority. The matrix plots use cases on two axes—risk (low/medium/high) and strategic value (low/medium/high)—and recommends actions for each quadrant.

High risk, high value: These are your most important AI initiatives. They deliver significant business value but carry substantial risk. Examples: automated underwriting, clinical decision support, supply chain optimization. **Action:** Invest heavily in guardrails. Assign top talent. Monitor closely. These systems are worth the effort.

High risk, low value: These use cases carry significant risk but do not deliver enough value to justify that risk. Examples: a highly automated customer-facing tool that improves efficiency marginally but exposes the organization to compliance violations. **Action:** Reconsider whether the use case is worth pursuing. If yes, add controls or reduce automation to lower risk. If no, deprioritize or cancel.

Low risk, high value: These are ideal use cases. They deliver value without significant risk. Examples: internal productivity tools, advisory systems, content generation for internal use. **Action:** Fast-track these. Apply lightweight governance. Let teams innovate quickly. These are your quick wins.

Low risk, low value: These use cases have little upside and little downside. Examples: experimental prototypes, niche tools with small user bases. **Action:** Tolerate these with minimal oversight. They are learning opportunities or minor conveniences. Do not over-invest in governance, but do not ignore it entirely—monitor for evolving risk.

Use the Risk-Control Matrix in portfolio reviews with your AI Steering Committee. It provides a visual way to discuss where the organization is over-investing (heavy governance on low-risk, low-value use cases) or under-investing (light governance on high-risk use cases). Adjust your portfolio mix to maximize value while managing risk.

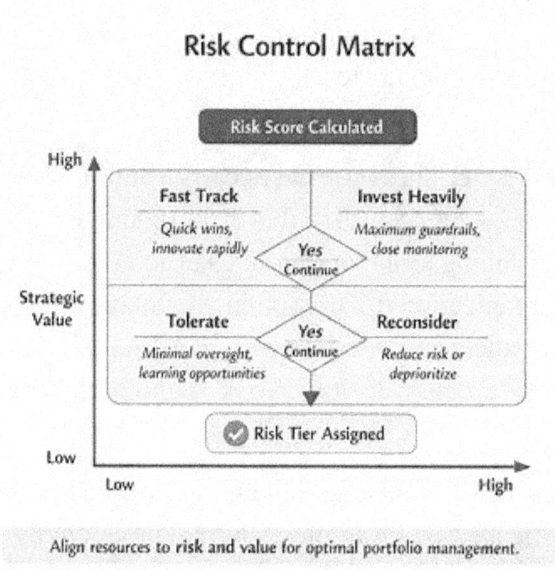

Align resources to risk and value for optimal portfolio management.

7.9 Dynamic Risk Reassessment

Risk is not static. Use cases evolve, threats change, and regulations are introduced. Dynamic risk reassessment means that you periodically re-score your AI portfolio and adjust governance as needed. This keeps your risk management aligned with reality.

Triggers for reassessment:

- **Scheduled reviews:** At least annually, re-score all active AI use cases.
- **Scope changes:** When a use case expands to new users, geographies, or features.
- **Incident or near-miss:** After any incident involving an AI system, reassess its risk.
- **Regulatory changes:** When new laws or rules are introduced that affect AI.
- **Model updates:** When models are retrained, updated, or replaced.
- **User feedback patterns:** If user complaints or overrides spike, reassess.

Reassessment process:

1. Review the original risk scores and documentation.
2. Identify what has changed (scope, data, model, regulations, environment).
3. Re-score relevant risk dimensions.
4. Calculate the new total score and determine if the tier has changed.
5. If the tier increases, implement additional governance requirements immediately.
6. If the tier decreased, consider relaxing governance (but document the justification).
7. Update documentation and notify relevant stakeholders.

Track risk evolution over time. If many use cases are moving from low to medium or medium to high, that signals that your AI program is becoming more complex and higher-stakes. You may need to increase governance capacity, invest in more sophisticated guardrails, or slow down deployment until controls catch up.

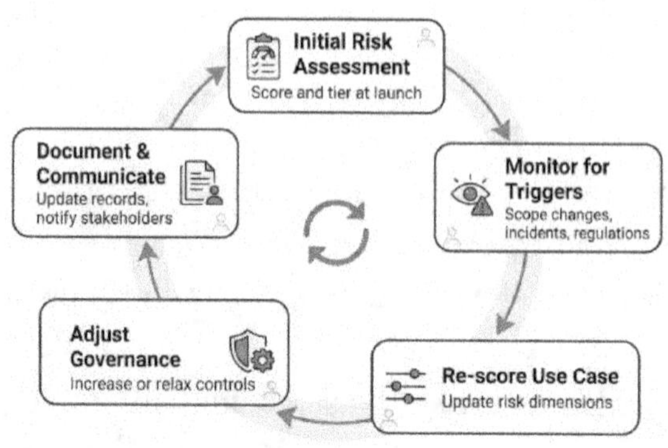

7.10   Playbook: Implementing Risk Triage

You have the framework, the scoring system, and the governance requirements. Now, how do you implement this in your organization? Here is a step-by-step playbook.

Step 1: Customize the scoring criteria. The framework provides a starting point, but you may need to adjust scoring criteria to fit your organization's context, risk appetite, and industry. Work with your AI Working Group to define what constitutes a "1," "2," or "3" for each dimension. Document these definitions clearly.

Step 2: Inventory your AI portfolio. List all active and planned AI use cases. For each, gather basic information:

description, owner, users, data sources, model type, and deployment status.

Step 3: Score each use case. For each use case, have the owner or a governance team member score the six risk dimensions. Calculate total scores and assign risk tiers. This is your initial risk classification.

Step 4: Define governance requirements for each tier. Use the guidance in Section 6.4 as a starting point, but customize it to your organization. What approvals are required? What testing? What monitoring? Document these requirements clearly and publish them so teams know what to expect.

Step 5: Validate scores and tiers. Have the AI Working Group review a sample of scores to ensure consistency and calibration. Discuss edge cases. Adjust scoring criteria if needed. This validation step prevents gaming or inconsistent application of the framework.

Step 6: Map current use cases to requirements. For each use case, compare its current governance to the requirements for its tier. Identify gaps. For example, a high-risk use case that has not had a fairness review is out of compliance. Create remediation plans for gaps.

Step 7: Integrate risk triage into the approval process. Make risk scoring a mandatory step in your AI use case proposal form (see Chapter 2). New use cases must be scored and tiered before they can be approved. This ensures the framework is applied consistently going forward.

Step 8: Establish reassessment cadence. Schedule annual risk reviews for all use cases. Also define triggers that require immediate reassessment (scope changes, incidents, etc.). Assign ownership for conducting reassessments.

Step 9: Monitor and iterate. Track how the framework is being used. Are teams scoring consistently? Are there frequent disagreements? Are high-risk use cases receiving appropriate oversight? Use feedback to refine the framework over time.

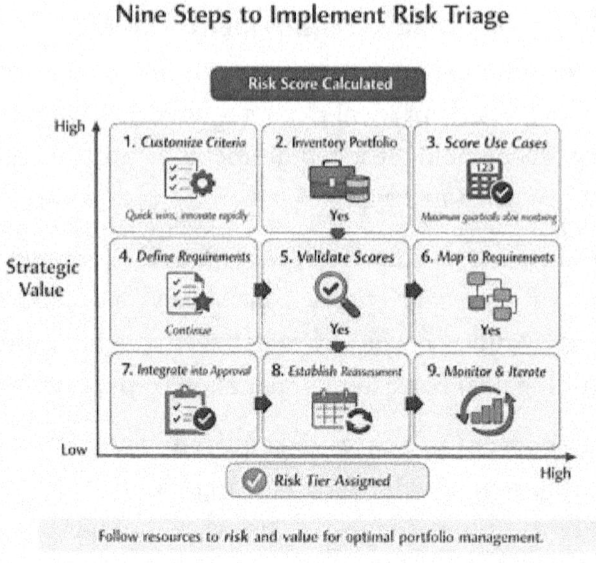

7.11 Common Pitfalls in Risk Assessment

Even with a structured framework, organizations make mistakes in risk assessment. Here are five common pitfalls and how to avoid them.

Pitfall 1: Scoring based on effort, not risk. Teams sometimes score use cases as "low risk" because they do not want to deal with the overhead of higher governance. **Avoidance:** Separate risk assessment from resource allocation. Risk scoring should be objective, based on impact and likelihood, not on convenience. If a use case is genuinely high risk but the business wants to move fast, the answer is to invest in appropriate guardrails, not to pretend the risk does not exist.

Pitfall 2: Ignoring cumulative risk. Each use case may be low risk, but collectively they create organizational risk if they all fail or misbehave simultaneously. **Avoidance:** Consider portfolio-level risk in addition to use-case-level risk. If you have 50 "low-risk" use cases all accessing the same sensitive data source, that data source itself becomes a high-risk dependency.

Pitfall 3: Static assessments that never update. Risk is assessed once at launch and never revisited, even as the use case evolves or the environment changes. **Avoidance:** Establish mandatory reassessment triggers and annual reviews. Make risk reassessment part of your operational rhythm.

Pitfall 4: Over-reliance on automation level. Teams assume that if a human is "in the loop," risk is automatically low. But if the human rubber-stamps AI recommendations without true review (automation bias), the oversight is illusory. **Avoidance:** Consider not just whether humans are involved, but how effective that involvement is. Monitor override rates and review quality.

Pitfall 5: Ignoring edge cases and rare events. Risk scoring often focuses on typical scenarios, missing rare but catastrophic events. **Avoidance:** Include "tail risk" in your assessment. Ask: "What is the worst-case scenario if this system fails? How likely is it? What would the consequences be?" For high-tail-risk use cases, apply higher governance even if routine operations seem safe.

7.12   Case: Applying Risk Triage to an AI Portfolio

Consider a healthcare organization with multiple AI use cases. Here is how they apply the Risk Triage Framework.

Use Case 1: Patient Appointment Reminder Bot

- **Scores:** Individual impact 1, organizational impact 1, automation 1 (advisory reminders), data sensitivity 2 (PII), model complexity 1 (simple rule-based), regulatory 2 (HIPAA)
- Total: 8 → Low risk
- **Governance:** Business owner approval, basic documentation, standard security checks, quarterly review.

Use Case 2: Radiology Image Flagging for Urgent Review

- **Scores:** Individual impact 2 (flags cases, but radiologist reviews), organizational 2 (moderate liability), automation 2 (flags for review), data sensitivity 3 (PHI, medical images), model complexity 3 (deep learning), regulatory 3 (HIPAA, FDA medical device rules)
- Total: 15 → High risk
- **Governance:** Steering Committee approval, comprehensive documentation, rigorous testing including fairness (across demographics), third-party validation, human-in-command (radiologist always reviews), continuous monitoring, semi-annual audits.

Use Case 3: Clinical Decision Support for Drug Interaction Warnings

- **Scores:** Individual impact 3 (affects patient safety), organizational 3 (high liability, reputational risk), automation 2 (provides warnings, physician decides), data

sensitivity 3 (PHI), model complexity 2 (rule-based with some ML), regulatory 3 (FDA, HIPAA)
- Total: 16 → High risk
- **Governance:** Steering Committee approval, legal and clinical expert review, extensive testing including clinical validation trials, human-in-the-loop (physician reviews all warnings), continuous monitoring, semi-annual audits.

Outcome: The organization focuses its limited resources on the two high-risk systems (radiology flagging and drug interaction warnings), ensuring they receive maximum scrutiny, testing, and oversight. The low-risk appointment bot receives lightweight governance, allowing the team to iterate and improve quickly without bureaucratic delays. This is proportional, effective risk management.

7.13   AI Reality Check

Before we close, let's address common myths about risk assessment:

- **Myth 1: "All AI is high risk."**
  Reality: Risk varies widely. Many AI use cases are low risk. Treating everything as high risk wastes resources and slows innovation.
- **Myth 2: "Risk assessment is a one-time checkbox."**
  Reality: Risk evolves. Reassess regularly and when circumstances change.
- **Myth 3: "Scoring is objective and eliminates judgment."**
  Reality: Scoring provides structure, but judgment is still required for edge cases, context, and trade-offs.

- **Myth 4: "Low risk means no governance."**
  Reality: Low risk means lightweight governance, not zero governance. All AI use cases need basic documentation, monitoring, and accountability.

Use this reality check to set expectations with your team and leadership. Risk-based governance is powerful, but it is not a magic formula. It requires discipline, judgment, and continuous attention.

7.14 Summary: Proportional Governance for AI

This chapter has shown you how to assess AI risk systematically, classify use cases into tiers, and apply proportional governance. You learned the six key risk dimensions—impact on individuals, impact on organization, automation level, data sensitivity, model complexity, and regulatory exposure—and how to score them to produce a total risk score. You learned the AI Risk Triage Framework, which maps scores to three tiers (low, medium, high) and defines governance requirements for each tier. You learned mitigation strategies to reduce risk, the Risk Control Matrix to align resources with risk and value, and the importance of dynamic reassessment as use cases evolve.

Risk-based governance is not about eliminating all risk. It is about understanding risk, prioritizing it, and managing it proportionally so you can innovate where it is safe to do so and exercise discipline where it is essential. The alternative—treating all AI the same—leads to either over-governance that stifles innovation or under-governance that exposes the organization to harm. Neither is acceptable. As an IT manager, your role is to implement and maintain an objective, transparent, and continuously improving risk triage process. When done right, risk triage enables your organization to scale AI safely, focusing guardrails where they matter most while keeping the path clear for low-risk innovation.

8 Monitoring, Drift, and Incident Response

8.1 When Silent Degradation Becomes Crisis

Your customer sentiment analysis tool has been running smoothly for eight months. It classifies support tickets as positive, neutral, or negative and routes negative cases to senior agents. No errors. No alerts. Performance dashboards show 92 percent accuracy—right where it has always been. Then a customer escalates a complaint to your CEO. She submitted three tickets about a billing error, all polite and factual, but the AI classified them as "positive" and routed them to a queue that was never monitored. She waited three weeks with no response. When your team investigates, they discover a subtle pattern: the model has drifted. Recent tickets use phrasing the model was not trained on—customers now say "I need help with" instead of "I have a problem with"—and the model interprets the word "help" as positive sentiment. Accuracy is still 92 percent overall, but it has dropped to 70 percent for a specific customer segment. No alerts fired. No dashboards flagged the issue. The system degraded silently until it became a crisis.

This chapter is about preventing that scenario. AI systems do not fail as traditional software does. They degrade quietly. They drift as data distributions shift. They develop emergent biases as user behavior changes. They perform well on average but fail catastrophically on edge cases. Monitoring, drift detection, and incident response are the guardrails that make these silent failures visible and manageable. You will learn how to continuously monitor AI system behavior, detect drift before it causes harm, and respond to incidents with speed and structure. By the end, you will have a practical incident playbook that your team can execute under pressure.

8.2 Why AI Systems Fail Quietly

Traditional software fails loudly. A server crashes. A database connection times out. An API returns an error code. These failures trigger alerts, logs, and automated recovery procedures. AI systems fail differently. They continue operating. Responses are generated. Dashboards show green. But the quality of decisions degrades. Outputs become less accurate, less fair, or less aligned with policy. The system appears to be working, even though it is actually broken.

There are three reasons why AI failures are silent: first, **probabilistic behavior**. AI systems produce outputs based on statistical patterns rather than deterministic rules. A model that is 90 percent accurate is wrong 10 percent of the time, but those errors may not trigger traditional monitoring because the system is still "functioning." You need monitoring that evaluates output quality, not just system availability.

Second, **data drift**. AI models are trained on historical data. Over time, the real-world data the model encounters changes. Customer language evolves. Market conditions shift. User behavior adapts. If the model was trained on data from 2023 and it is now 2026, its understanding of the world may be outdated. Traditional software does not degrade this way—code does not "forget" how to execute a loop—but models do. They become less accurate as the gap between training data and live data widens.

Third, **emergent behavior**: AI systems can develop unexpected actions through learning from interactions and feedback. These behaviors may be helpful or harmful and often go unnoticed until an issue arises, when they can be harmful.

For IT managers, actively overseeing AI systems should be a fundamental practice rather than a secondary concern. Monitoring needs to be constant, multifaceted, and seamlessly built into daily operations. It's essential to ensure not only that the technology is functioning, but also that its outcomes remain accurate, fair, safe, and in line with policy requirements.

8.3 The Three Layers of AI Monitoring

Effective AI monitoring operates at three layers: **system health, model performance, and business outcome**. Each layer provides different signals and requires different tools.

System health monitoring tracks the infrastructure and technical operation of the AI system: API response times, error rates, throughput, resource utilization (CPU, memory, GPU), request volumes, and uptime. These are familiar metrics from traditional IT operations. They tell you whether the system is available and responsive, but they do not tell you whether it is producing good outputs. A system can have 99.9 percent uptime and still be generating biased or inaccurate results.

Model performance monitoring tracks the quality of the model's predictions or outputs: accuracy, precision, recall, F1 score, confidence distributions, false positive and false negative rates, and fairness metrics (demographic parity, equalized odds). For generative models, monitor output coherence, hallucination rates, toxicity scores, and policy compliance. Model performance metrics indicate whether the AI is performing its job correctly. This is the most important layer for detecting drift, bias, and degradation.

Business outcome monitoring tracks whether the AI system is achieving its intended business goals: customer satisfaction scores, resolution times, conversion rates, cost savings, user complaints, override rates (how often humans reverse AI decisions), and operational metrics specific to the use case. Business outcome metrics tell you whether the AI is delivering value. A model can perform technically well but still fail to achieve business goals if the use case is poorly designed or user trust is low.

All three layers are necessary. System health monitoring catches infrastructure failures. Model performance monitoring catches drift and quality degradation. Business outcome monitoring catches misalignment between technical performance and real-world impact. Integrate all three into your monitoring dashboards and alerting systems.

Three Layers of AI Monitoring

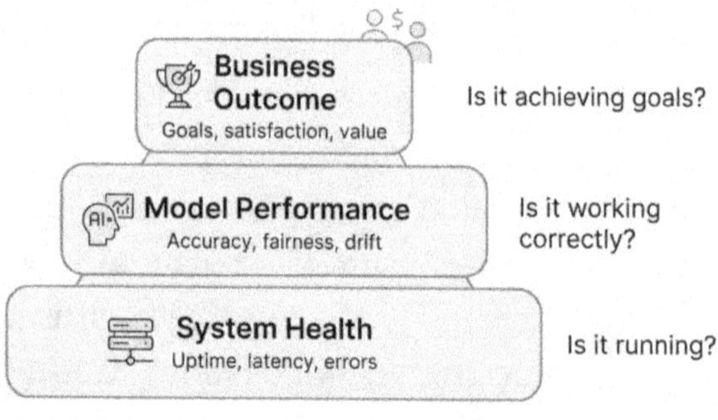

8.4 Detecting Data Drift and Concept Drift

Drift is the gradual change in data distributions or relationships between inputs and outputs over time. There are two types: **data drift** (the distribution of input features changes) and **concept drift** (the relationship between inputs and outputs changes). Both can degrade model performance silently.

Data drift occurs when the characteristics of incoming data change. For example, in a loan approval model, if the average applicant age shifts from 35 to 45, or if income distributions change due to economic shifts, the model may struggle because it was trained on a different demographic profile. Data drift is common in dynamic environments: customer behavior evolves, economic conditions shift, seasonal patterns emerge, and product offerings change.

Concept drift occurs when the relationship between features and outcomes changes over time. For example, in fraud detection, fraudsters adapt their tactics. Behaviors that once indicated fraud may become normal, and new fraud patterns emerge. The model's learned rules become outdated. Concept drift is especially problematic because it cannot be detected solely by monitoring input distributions. You must monitor prediction accuracy against ground truth.

Detecting data drift: Use statistical methods to compare the distribution of live data with that of training data. Common techniques include: comparing feature means, standard deviations, and quantiles over time; using statistical tests (Kolmogorov-Smirnov test, chi-square test) to detect significant distribution shifts; and tracking feature importance—if a feature that was important during training becomes irrelevant, or vice versa, that is a signal of drift—set thresholds for acceptable drift and trigger alerts when thresholds are exceeded.

Detecting concept drift: Monitor model performance metrics against ground truth. For supervised learning, if you have labeled data, compare predictions to actual outcomes over time. If accuracy drops significantly, concept drift is likely. For unsupervised or generative models, use proxy metrics such as user feedback (thumbs up/down, complaints), human override rates, or periodic manual review of outputs. Also monitor confidence distributions: if the model's confidence is declining (more predictions near the decision boundary), that may indicate uncertainty due to concept drift.

Responding to drift: When drift is detected, you have several options: retrain the model on recent data to update its understanding; adjust decision thresholds or calibration to account for distribution changes; add new features that capture emerging patterns; or pause the system and conduct a full review if drift is severe. Do not ignore drift signals. They are early warnings that the model is losing alignment with reality.

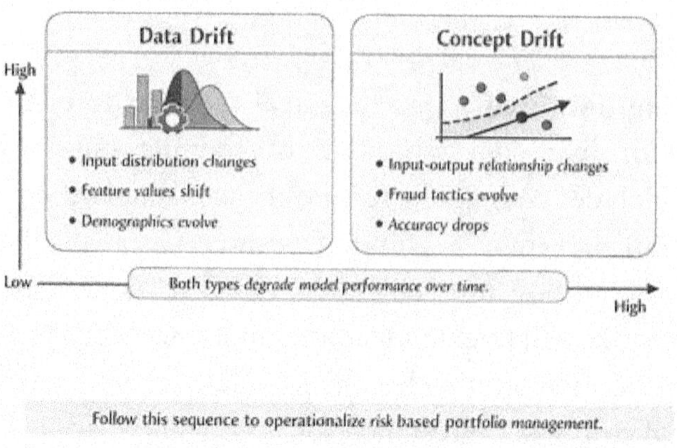

Follow this sequence to operationalize risk based portfolio management.

8.5 Monitoring Fairness and Bias Over Time

Fairness is not a one-time test. Even if a model passes fairness checks at launch, bias can emerge or worsen over time due to data drift, changes in user behavior, or feedback loops. Continuous fairness monitoring ensures that models remain equitable throughout their operational lifecycle.

Establish fairness baselines. At launch, measure fairness metrics across demographic groups: demographic parity, equalized odds, predictive parity. Document these baselines. For example, if a hiring tool has a selection rate of 30 percent for Group A and 28 percent for Group B at launch, that 2 percentage-point gap is your baseline.

Monitor fairness metrics continuously. Re-measure fairness metrics regularly (monthly or quarterly, depending on use-case criticality and data volume). Track whether gaps are widening, narrowing, or stable. If a gap that was 2 percentage points at launch grows to 10 percentage points six months later, investigate immediately.

Detect emergent bias. Bias can emerge from feedback loops. For example, if an AI system recommends content, and users engage more with certain types of content, the model learns to recommend similar content, creating a reinforcing loop. Over time, the system may become biased toward a narrow set of preferences, excluding diverse viewpoints. Monitor not only model outputs but also downstream effects: Are certain user groups being systematically disadvantaged? Are certain outcomes becoming more homogeneous?

Segment performance by group. Do not rely solely on aggregate metrics. A model with 90 percent overall accuracy might have 95 percent accuracy for Group A and 80 percent accuracy for Group B. Segment performance metrics by protected characteristics (where legally permissible and relevant) to identify disparities. Track false positive and false negative rates, not just overall accuracy.

Respond to fairness degradation. If fairness metrics degrade, take immediate action: investigate root causes (is the training data becoming less representative? Are certain features driving bias?), retrain with rebalanced data or fairness constraints, adjust decision thresholds for different groups if justifiable, or add human review for affected groups. Document all decisions and rationales.

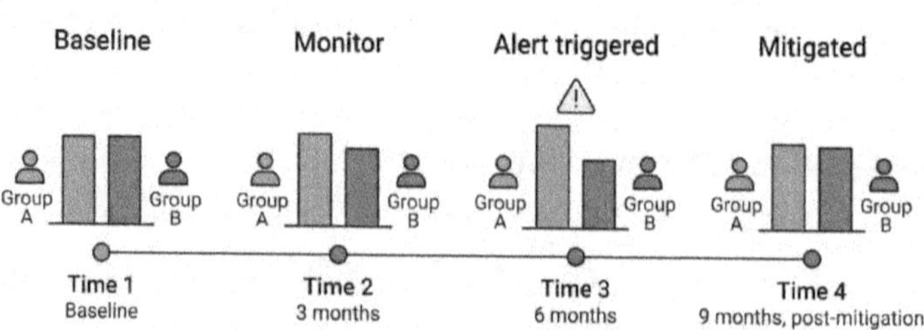

8.6 Alerting Strategies: When and How to Alert

Monitoring generates data. Alerting turns that data into action. Effective alerting ensures that critical issues are surfaced quickly without overwhelming teams with false alarms. Poorly designed alerting leads to alert fatigue, where teams ignore notifications because most are false positives or low priority.

Tiered alerting assigns severity levels to alerts based on impact and urgency. Define four tiers:

- **Critical:** Immediate action required within minutes. Examples: system outage, model accuracy drops below safety threshold, data breach detected, fairness violation exceeds acceptable bounds. Route to on-call engineers, AI governance lead, and operations manager. Use SMS, phone, and Slack/Teams notifications.
- **High:** Action required within hours. Examples: drift detected above threshold, error rate spike, user complaints spike, performance degradation in specific segment. Route

146

to system owner, data science team, and governance team. Use email and Slack/Teams.

- **Medium:** Action required within days. Examples: gradual performance decline, minor drift, configuration warnings. Route to system owner and relevant teams. Use email and weekly digest reports.
- **Low:** Informational or action required within weeks. Examples: usage pattern changes, minor anomalies, non-critical recommendations. Route to system owner. Use weekly or monthly digest reports.

Threshold-based alerting triggers alerts when metrics cross predefined thresholds. For example: alert if model accuracy drops below 85 percent, if false positive rate exceeds 10 percent, if response time exceeds 2 seconds, or if daily request volume drops by more than 50 percent. Set thresholds based on baselines, business requirements, and risk tolerance. Review and adjust thresholds quarterly.

Anomaly detection alerting uses statistical or machine learning techniques to detect unusual patterns that may not cross fixed thresholds. For example, if request volume suddenly spikes at an unusual time, or if a feature distribution shifts abruptly, or if user sentiment drops sharply, trigger an alert. Anomaly detection catches unexpected issues that threshold alerts might miss.

Alert routing and escalation: Define clear routing rules. Who receives which alerts? For critical alerts, use escalation paths: if the primary on-call does not acknowledge within 10 minutes, escalate to secondary on-call and manager. For high and medium alerts, route to relevant teams based on issue type (data science, engineering, governance). For low-priority alerts, aggregate into digest reports to reduce noise.

Alert fatigue prevention: Monitor alert frequency and resolution rates. If teams are receiving more than 10 alerts per day, or if most alerts result in no action, you have alert fatigue. Address by: tightening alert criteria (reduce false positives), consolidating related alerts (group similar issues), tuning thresholds to reduce noise, or automating responses for routine issues.

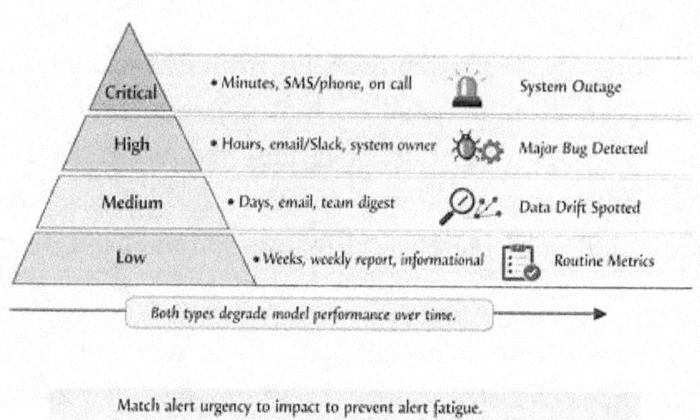

8.7 Incident Detection and Classification

An incident is an event where an AI system causes or risks causing harm: incorrect decisions, unfair outcomes, data leakage, policy violations, safety failures, or reputational damage. Incident detection and classification are the first steps in response. The faster you detect and classify incidents, the faster you can contain and resolve them.

Incident detection sources include: automated alerts (system monitoring, model performance, anomaly detection), user complaints (support tickets, social media, escalations), internal reports (employees noticing issues), audit findings (periodic reviews uncovering problems), and regulatory inquiries (external parties raising concerns). Establish clear channels for reporting incidents from all sources.

Incident classification assigns severity and type to each incident. Use a four-tier severity model:

- **Severity 1 (Critical):** Incident causes or risks imminent harm to individuals, significant organizational damage, or legal/regulatory violations. Examples: AI system denies life-critical service inappropriately, data breach exposes sensitive information, model exhibits severe discriminatory bias.
- **Severity 2 (High):** Incident causes moderate harm or significant operational disruption. Examples: model accuracy drops causing customer dissatisfaction, fairness violation exceeds thresholds but not catastrophic, system downtime affecting business operations.
- **Severity 3 (Medium):** Incident causes minor harm or inconvenience. Examples: isolated incorrect predictions, minor policy violations, degraded user experience.
- **Severity 4 (Low):** Incident is informational or has minimal impact. Examples: edge case anomaly, minor configuration issue, non-critical bug.

Incident types categorize the nature of the incident: performance degradation, bias or fairness issue, security or privacy breach, policy violation, safety failure, or reputational harm. Classifying by type helps route incidents to the appropriate response team (data science, security, legal, governance) and informs root cause analysis.

Initial triage: When an incident is detected, perform rapid triage within minutes: confirm the incident is real (not a false alarm), assign severity and type, identify affected systems and users, determine immediate containment actions (pause system, roll back, route to human review), and notify relevant stakeholders. Document all triage decisions.

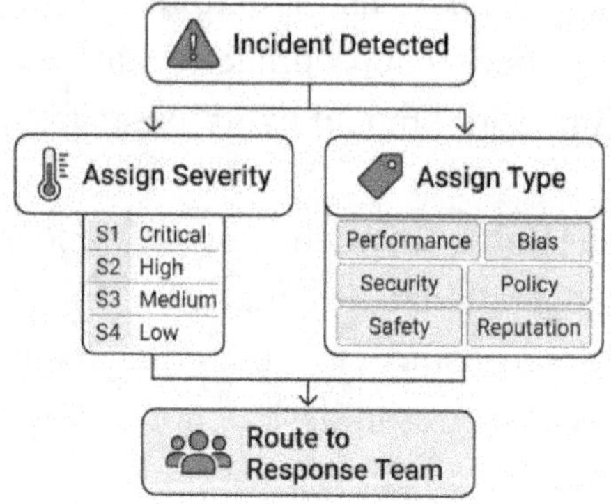

8.8 The AI Incident Response Playbook

An incident response playbook is a documented, step-by-step process for responding to AI incidents. It defines roles, responsibilities, decision points, and communication protocols so teams can act decisively under pressure. The playbook should be tailored to your organization but follow a standard structure.

Phase 1: Detect and Triage (0–15 minutes)

- Confirm incident is real and not false alarm.
- Assign severity and type.
- Identify affected systems, users, and data.
- Notify incident commander (designated role for coordinating response).
- Assess immediate risks and decide on containment actions.

Phase 2: Contain (15–60 minutes)

- Implement containment actions to prevent further harm: pause or disable the AI system, roll back to previous model version, route all decisions to human review, isolate affected data or systems, or block specific inputs or outputs.
- Notify stakeholders: system owner, AI governance lead, legal, communications, affected business units.
- Document all containment actions taken and their outcomes.

Phase 3: Investigate (1–48 hours, depending on severity)

- Assemble investigation team: data science, engineering, governance, domain experts.
- Analyze root cause: what caused the incident? Data drift? Model bug? Policy violation? External attack?
- Review logs, model behavior, data changes, and user reports.
- Determine scope: how many users or decisions were affected? What is the potential harm?
- Document findings in incident report.

Phase 4: Remediate (hours to weeks, depending on severity)

- Develop remediation plan: retrain model, fix data pipeline, update guardrails, revise policy, enhance monitoring.
- Implement fixes and test thoroughly before redeployment.

- Validate that remediation addresses root cause and does not introduce new issues.
- Communicate remediation plan to stakeholders.

Phase 5: Communicate (ongoing)

- Internal communication: keep leadership, affected teams, and incident responders informed.
- External communication (if warranted): notify affected users, regulators, or the public. Work with legal and communications teams on messaging.
- Provide transparency: explain what happened, what harm occurred, and what actions were taken.

Phase 6: Learn and Improve (post-incident)

- Conduct post-incident review (PIR): what went well? what went poorly? what can be improved?
- Update monitoring, alerting, and guardrails based on lessons learned.
- Revise incident response playbook if gaps were identified.
- Share findings with broader AI governance team to prevent similar incidents.

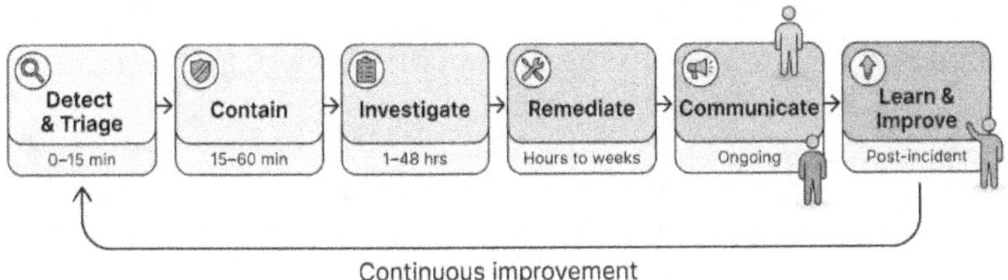

Follow a structured process to manage incidents under pressure.

8.9 Roles and Responsibilities in Incident Response

Effective incident response requires clear roles. Ambiguity leads to delays, duplicated effort, and gaps in response. Define these roles in advance and practice them through tabletop exercises.

Incident Commander: Single point of authority who coordinates response, makes decisions, and ensures communication. Typically, a senior IT manager, AI governance lead, or designated on-call leader. Responsibilities: triage incident, declare severity, activate response team, make containment decisions, coordinate across teams, communicate with leadership, and ensure documentation.

Technical Lead: Senior engineer or data scientist responsible for technical investigation and remediation. Responsibilities: analyze root cause, review logs and model behavior, develop technical fix, coordinate with engineering and data science teams, and validate remediation.

Governance Lead: AI governance or compliance representative who ensures incident response aligns with policies and regulations. Responsibilities: assess policy and regulatory implications, advise on communication and disclosure requirements, coordinate with legal and compliance, and document governance aspects.

Communications Lead: Representative from communications or public relations who manages external and internal messaging. Responsibilities: draft communications, coordinate with legal on disclosures, manage media inquiries, and ensure consistent messaging.

Business Owner: Representative from the affected business unit who provides domain context and assesses business impact. Responsibilities: explain use case context, assess customer or user impact, prioritize remediation based on business needs, and communicate with affected users.

Subject Matter Experts: Domain experts, security specialists, or external consultants who provide specialized knowledge. Responsibilities: advise on specific aspects (clinical safety, security vulnerabilities, fairness), review findings, and recommend specialized solutions.

Document these roles in your incident response playbook. Include contact information, escalation paths, and decision authority. Ensure everyone knows their role before an incident occurs.

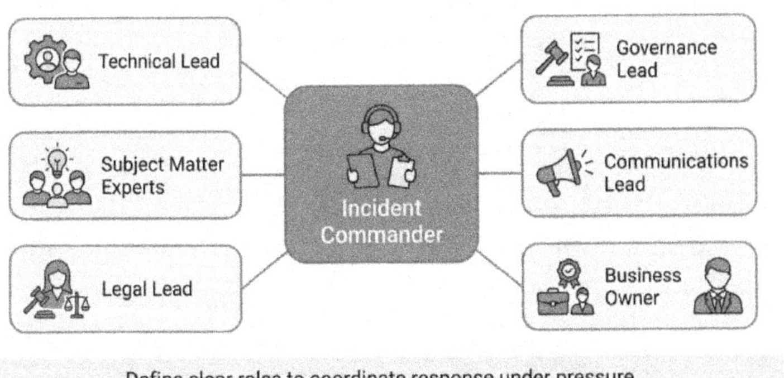

Define clear roles to coordinate response under pressure.

8.10   9 Communication During and After Incidents

Communication is often the weakest link in incident response. Poor communication creates confusion, erodes trust, and amplifies harm. Effective communication requires planning, clarity, and transparency.

Internal communication keeps leadership, affected teams, and responders informed. Use a dedicated incident channel (Slack, Teams) for real-time updates. Provide regular status updates (every 30–60 minutes for critical incidents), including: current situation, actions taken, next steps, and estimated resolution time. Avoid jargon. Use plain language so non-technical stakeholders can understand.

External communication may be required if users, customers, regulators, or the public are affected. Coordinate with legal, compliance, and communications teams before making external statements. Key principles: acknowledge the incident promptly, explain what happened in accessible terms (avoid technical jargon), describe the potential harm or impact, outline what actions were taken to contain and remediate, provide a point of contact for questions or concerns, and commit to transparency and updates as more is learned.

Regulatory communication is mandatory in some cases (data breaches, safety incidents, violations of regulated services). Know your obligations: which regulators must be notified? What are the timeframes (e.g., 72 hours for GDPR breaches)? What information must be disclosed? Work with legal and compliance to ensure timely, accurate reporting.

Post-incident communication shares lessons learned. Internally, conduct a post-incident review (PIR) meeting with all responders. Discuss what went well, what went poorly, and what should change. Document findings and action items. Share the PIR summary with AI governance and leadership. Externally (if appropriate), publish a transparent post-mortem: what happened, why it happened, what was fixed, and what is being done to prevent recurrence. Transparency builds trust.

Avoid these communication pitfalls: delays (waiting too long to communicate creates speculation and distrust), minimization (downplaying the incident makes it worse when full scope emerges), blame (focus on systems and processes, not individuals), and silence (lack of communication suggests indifference or cover-up).

8.11   Case Example: Responding to a Fairness Incident

To make incident response concrete, consider a case example: a hiring tool exhibits unexpected bias.

Detection: The system has been live for six months. Routine monthly fairness monitoring reveals that the selection rate for women has dropped from 32 percent (baseline) to 18 percent. An automated alert is triggered (fairness gap exceeds 5 percentage-point threshold). Incident is detected.

Triage (0–15 minutes): Incident Commander reviews alert. Confirms it is real (not a data artifact). Assigns **Severity 2 (High)** because it affects a protected characteristic and could lead to legal liability. Type: **Bias/Fairness**. Notifies AI governance lead, Technical Lead, and legal.

Contain (15–60 minutes): Incident Commander pauses the hiring tool. All new applications are routed to manual review by HR staff. The affected business unit is notified. Legal is briefed. No external communication yet (investigating first).

Investigate (1–24 hours): Technical Lead assembles data science team. They analyze recent data and discover that a new feature was added to the model two months ago (time-to-hire in prior roles). This feature correlates with gender because women in the dataset had a longer time to hire (often due to career breaks). The model learned to penalize a longer time-to-hire, indirectly penalizing women. Root cause identified: proxy bias introduced by the new feature.

Remediate (24–72 hours): Technical Lead develops a remediation plan: remove the problematic feature, retrain the model on balanced data, add fairness constraints during training, and conduct red team testing for similar proxy biases. The model is retrained and tested. Fairness metrics return to baseline. Legal reviews and approves redeployment.

Communicate (ongoing): Internally, HR and leadership are briefed. Externally, the company proactively notifies applicants who were processed during the biased period, offering a-review of their applications. Regulatory reporting is not required (no legal violation), but proactive transparency is chosen. Communications team drafts messaging.

Learn and Improve (post-incident): Post-incident review identifies improvements: add fairness review to the feature addition process, increase fairness monitoring frequency from monthly to weekly for high-risk models, and enhance red-team testing to catch proxy bias earlier. Playbook is updated.

Outcome: Incident is contained within 1 hour, remediated within 72 hours, and communicated transparently. Trust is maintained. Applicants appreciate proactive notification. Lessons learned prevent similar incidents. This is an effective incident response.

8.12   Building an Incident Response Capability

Having a playbook is not enough. You must build an operational capability to execute it. This requires training, practice, tools, and culture.

Training: Conduct incident response training for all roles. Use workshops, tabletop exercises, and simulations. Walk through the playbook step by step. Discuss real-world scenarios and decision points. Ensure everyone knows their responsibilities.

Tabletop exercises: Simulate incidents without actually triggering them. Present a hypothetical scenario: "An automated loan system has denied 200 applications due to a data pipeline error.

Applicants are complaining on social media. Go." Have the team walk through detection, triage, containment, investigation, and communication. Observe where confusion or delays occur. Refine the playbook based on findings.

Tools and infrastructure: Ensure you have the tools needed to respond: incident tracking system (Jira, ServiceNow), dedicated communication channels (Slack, Teams), logging and monitoring dashboards, rollback and deployment tools, contact lists and escalation paths, and templates for communication (internal updates, external statements, regulatory reports).

Culture: Foster a culture where incidents are learning opportunities, not failures. Encourage the reporting of near misses and anomalies. Reward teams for catching issues early. Conduct blameless post-incident reviews that focus on systems and processes, not individuals. Share lessons learned openly to improve organizational capability.

Continuous improvement: After every incident or exercise, update the playbook. Add new scenarios. Refine decision criteria. Update contact lists. Incident response is a living capability that improves through practice and learning.

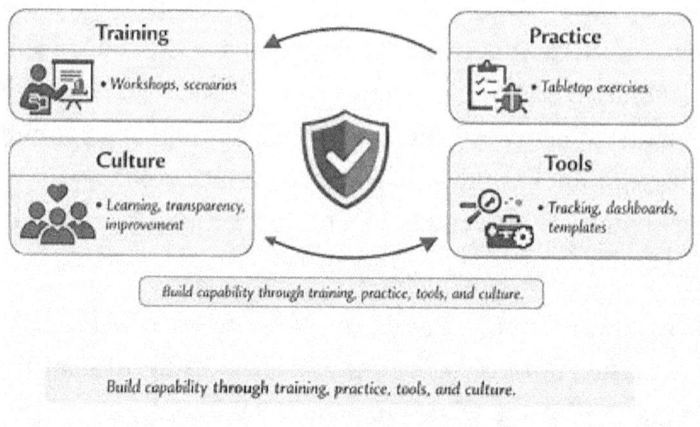

Build capability through training, practice, tools, and culture.

8.13 Playbook: Establishing Monitoring and Incidents

You have learned the principles, strategies, and processes. Now, how do you implement monitoring and incident response in your organization? Here is a practical playbook.

Step 1: Define monitoring requirements. For each AI use case, determine: What system health metrics should be tracked? What model performance metrics? What fairness metrics? What business outcome metrics? What alert thresholds? Document these requirements in a monitoring plan.

Step 2: Implement monitoring infrastructure. Implement solutions for logging, metrics collection, dashboards, and alerting. Where feasible, integrate these tools with existing observability platforms. Ensure that data science, engineering, and governance teams have appropriate access to dashboards.

Step 3: Establish baselines. Gather baseline metrics for each use case during the initial 1–3 months of operation. Record these baselines for accuracy, fairness, performance, and business results. These benchmarks serve as references to identify any drift or decline over time.

Step 4: Configure alerts. Set up tiered alerts (critical, high, medium, low) with appropriate thresholds, routing, and escalation paths. Test alerts to ensure they fire correctly. Tune thresholds to minimize false positives.

Step 5: Draft incident response playbook. Customize the playbook template from this chapter. Define roles, phases, decision criteria, communication protocols, and contact lists. Review with stakeholders and iterate.

Step 6: Train and practice. Conduct incident response training for all roles. Run at least one tabletop exercise per quarter. Simulate different incident scenarios (bias, drift, security, safety).

Step 7: Integrate into operations. Make monitoring and incident response part of daily operations. Assign on-call rotations. Review dashboards in team meetings. Conduct monthly performance reviews that include monitoring metrics.

Step 8: Continuously improve. After every incident or exercise, conduct post-incident reviews. Update playbook, monitoring plans, and alert configurations based on lessons learned. Share findings with the AI governance team.

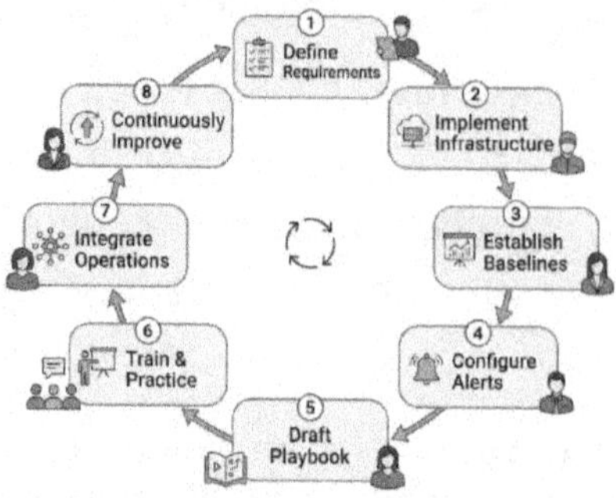

Build monitoring and incident response into your operational rhythm.

8.14   AI Reality Check

Before we close, let's address common myths about monitoring and incident response:

- **Myth 1: "Our AI system has been stable for months, so we don't need intensive monitoring."**
  Reality: Stability does not mean immunity. Drift and degradation can happen suddenly. Continuous monitoring is the only way to detect issues before they become crises.

- **Myth 2: "If accuracy is stable, the system is fine."**
  Reality: Aggregate accuracy can mask problems. Segment performance by group, context, and edge cases. A model with stable overall accuracy can still fail in specific segments.

- **Myth 3: "Incident response is for security teams, not AI teams."**
  Reality: AI incidents are distinct from traditional IT incidents. They require domain expertise, data science

knowledge, and an understanding of fairness and ethics. AI teams must own incident response.

- **Myth 4: "We'll figure out incident response if something happens."**
  Reality: Incidents create pressure, confusion, and urgency. Without a pre-defined playbook, the response will be chaotic. Prepare before incidents occur.

Use this reality check to set expectations and build urgency for monitoring and incident response capabilities.

8.15 Summary: From Detection to Resolution

This chapter has shown you how to maintain operational assurance for AI systems over time. You learned why AI systems fail quietly, the three layers of monitoring (system health, model performance, business outcome), how to detect data drift and concept drift, how to monitor fairness continuously, effective alerting strategies, incident detection and classification, the AI incident response playbook, roles and responsibilities, communication during and after incidents, and how to build an incident response capability through training, practice, and culture.

Monitoring and incident response are not optional extras. They are essential guardrails for AI at scale. AI systems are not fire-and-forget. They require continuous observation, calibration, and readiness to respond when things go wrong. The organizations that succeed with AI will be those that detect issues early, respond decisively, and learn from every incident. As an IT manager, your role is to build and maintain this capability, ensuring that your team has the tools, training, and playbooks needed to keep AI systems aligned with mission, safe, and trustworthy—even under pressure.

9 Legal, Compliance, and Regulatory Readiness

9.1 When Compliance Catches You Unprepared

Your customer service AI has been running successfully in three countries for 18 months. It routes inquiries, suggests responses, and automates simple requests. Then your legal team forwards an email from the European Data Protection Board. A customer filed a complaint under the EU AI Act, which went into effect six months ago. Your system is classified as "high-risk" because it makes decisions affecting access to services. Under the Act, you must provide transparency about how decisions are made, maintain technical documentation, conduct conformity assessments, and allow customers to challenge automated decisions. Your team scrambles. No one knew the AI Act applied. You have no technical documentation in the required format. You cannot explain how specific decisions were made. The conformity assessment will take months. Legal estimates fines could reach 30 million euros or six percent of global revenue. Operations wants to shut down the system immediately. The business is furious—this tool drives 40 percent of support efficiency. This is what happens when regulatory readiness is an afterthought.

AI regulation is no longer hypothetical. Laws are being enacted across jurisdictions, covering everything from facial recognition to hiring tools to healthcare AI. These laws impose real obligations: documentation, testing, audits, transparency, appeals, and liability. Non-compliance carries serious consequences: fines, enforcement actions, reputational damage, and operational shutdowns. This chapter shows you how to build regulatory readiness into your AI program from the start. You will learn the major regulatory frameworks, sector-specific requirements, how to design compliance once and localize it as laws evolve, and how to maintain readiness as your portfolio scales. The goal is not to become a legal expert, but to ensure your AI systems can operate confidently in any jurisdiction without last-minute scrambles.

9.2 Regulatory Landscape: Why Now, Why Global

AI regulation has arrived. For years, AI operated in a gray zone—encouraged by governments as innovation, loosely governed by existing laws, but lacking specific rules. That era is over. Jurisdictions worldwide are enacting AI-specific laws driven by three forces: high-profile AI failures (bias in hiring, wrongful arrests from facial recognition, discriminatory lending), recognition that existing laws are insufficient (data protection laws do not address algorithmic accountability), and geopolitical competition (nations want to shape global AI norms).

The regulatory landscape is fragmented. There is no single global AI law. Instead, there are overlapping frameworks: horizontal regulations that apply across sectors (the EU AI Act, China's algorithm regulations), sector-specific regulations (FDA rules for medical AI, financial services regulations for credit models), and data protection laws with AI implications (GDPR, CCPA, HIPAA). Organizations operating in multiple jurisdictions face a patchwork of compliance requirements: what is legal in one country may be restricted or banned in another.

For IT managers, this creates operational complexity. You cannot assume that a system deployed in the United States will comply with EU rules, or that a system designed for financial services will meet healthcare standards. You must design compliance with global portability: a foundational set of controls that satisfies the strictest requirements, which can then be localized for specific jurisdictions and sectors. This approach—compliance once, localize as needed—is more efficient than building separate systems for each jurisdiction.

The stakes are significant. Regulatory violations carry financial penalties (GDPR fines up to 4 percent of global revenue, EU AI Act fines up to 6 percent), operational consequences (regulators can ban non-compliant systems), reputational harm (public enforcement actions damage brand), and legal liability (individuals harmed by AI can sue). Regulatory readiness is not optional. It is an operational requirement for AI at scale.

9.3 The EU AI Act: The Global Benchmark

The EU AI Act is the world's first comprehensive AI regulation and sets the benchmark that other jurisdictions are likely to follow. It uses a risk-based approach, categorizing AI systems into four tiers: unacceptable risk (banned), high risk (strict requirements), limited risk (transparency obligations), and minimal risk (no specific regulation). Understanding the AI Act provides a template for compliance across the board.

Unacceptable risk (banned): AI systems that pose unacceptable threats to safety, livelihoods, or rights are prohibited. Examples include: social scoring by governments, real-time biometric identification in public spaces (with narrow exceptions), subliminal manipulation, and exploiting vulnerabilities of children or disabled persons. If your use case falls into this category, you cannot deploy it in the EU, regardless of safeguards.

High risk (strict requirements): AI systems that significantly affect safety or fundamental rights are subject to comprehensive obligations. Examples include: biometric identification and categorization, critical infrastructure management (energy, water, transport), educational and vocational training tools, employment and worker management (hiring, promotion, monitoring), access to essential services (credit scoring, benefits eligibility, emergency response), law enforcement tools (evidence evaluation, case prioritization), migration and border control, and judicial decision support. For high-risk systems, you must: maintain technical documentation, implement quality management systems, conduct conformity assessments (third-party audits in some cases), register the system in an EU database, ensure human oversight, provide transparency and explainability, maintain accuracy and robustness, implement cybersecurity measures, and establish post-market monitoring.

Limited risk (transparency obligations): AI systems that interact with humans or generate content must disclose that users are interacting with AI. Examples include: chatbots, deepfakes, and AI-generated content. You must provide clear, conspicuous notice that the user is interacting with AI and label synthetic content as AI-generated.

Minimal risk (no specific regulation): Most AI systems fall into this category: recommender systems, spam filters, inventory management, non-sensitive analytics. These face no specific obligations under the AI Act but remain subject to general data protection and consumer protection laws.

Compliance timeline: The EU AI Act entered into force in August 2024. Bans on unacceptable risk systems took effect immediately. High-risk requirements phase in over 24–36 months, with full compliance required by mid-2026. Organizations must assess their portfolios and ensure high-risk systems meet requirements before deadlines.

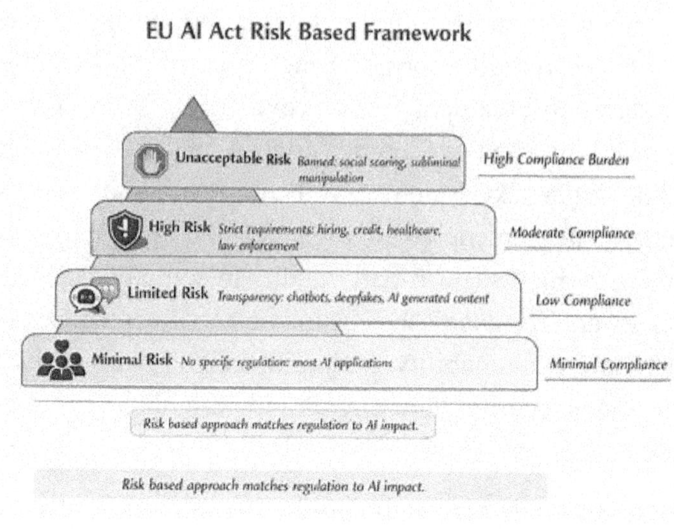

9.4 Regulations: Healthcare, Finance, and Beyond

In addition to horizontal AI laws such as the EU AI Act, sector-specific regulations impose additional requirements. If your AI operates in healthcare, financial services, or other regulated sectors, you must comply with both horizontal and sector rules.

Healthcare and medical AI: Medical AI is heavily regulated due to patient safety. In the United States, the FDA regulates AI as a medical device if it diagnoses, treats, or prevents disease. You must obtain FDA clearance or approval before deployment, submit clinical validation data, implement quality management systems (ISO 13485), maintain detailed documentation, and report adverse events. In the EU, medical AI is regulated under the Medical Device Regulation (MDR) and In Vitro Diagnostic Regulation (IVDR), with similar requirements. Key obligations: clinical evidence of safety and efficacy, risk management processes, post-market surveillance, and compliance with harmonized standards. Even non-medical healthcare AI (appointment scheduling, administrative tools) must comply with HIPAA (US) or GDPR (EU) to protect patient data.

Financial services AI: Credit scoring, loan decisioning, fraud detection, and investment advice tools are regulated by financial authorities. In the US, the Equal Credit Opportunity Act (ECOA) and Fair Credit Reporting Act (FCRA) prohibit discrimination and require transparency. Adverse action notices must explain why credit was denied. The Consumer Financial Protection Bureau (CFPB) enforces these laws and has issued guidance on AI in lending. In the EU, financial AI must comply with the GDPR, the AI Act, and sector-specific rules such as the Markets in Financial Instruments Directive (MiFID II). Key obligations: explainability (customers must understand the decisions), fairness testing (no discrimination based on protected characteristics), model validation and testing, audit trails of decisions, and human review for adverse decisions.

Employment AI: Hiring, promotion, and performance evaluation tools are subject to anti-discrimination laws. In the US, Title VII of the Civil Rights Act prohibits employment discrimination and is enforced by the Equal Employment Opportunity Commission (EEOC). The EEOC has issued guidance stating that AI tools can violate civil rights laws if they produce discriminatory outcomes, even without discriminatory intent. In the EU, employment AI is classified as high-risk under the AI Act. In New York City, Local Law 144 requires bias audits for automated employment decision tools. Key obligations: bias testing and audits, transparency to applicants, human oversight, and documentation of validation methods.

Other sectors: Education AI (student evaluation, admissions) faces privacy and equity requirements. Government AI (benefits eligibility, law enforcement) faces transparency, due process, and accountability requirements. Autonomous vehicles face transportation safety regulations. Each sector imposes unique

Know your sector's specific AI compliance requirements.

obligations that layer onto horizontal AI laws.

9.5 Protection and Privacy: GDPR, HIPAA, etc

AI systems process data—often large volumes of personal, sensitive data. Data protection and privacy laws regulate how that data is collected, used, stored, and shared. These laws predate AI but have significant implications for AI systems.

GDPR (EU General Data Protection Regulation): GDPR governs the processing of personal data in the EU and applies to any organization processing the data of EU residents. Key requirements for AI: lawful basis for processing (consent, contract, legitimate interest, legal obligation), data minimization (collect only what is necessary), purpose limitation (use data only for stated purposes), transparency (inform individuals how data is used), individual rights (access, correction, deletion, portability), and automated decision-making rules (individuals have the right not to be subject to solely automated decisions with significant effects, unless exceptions apply). For AI, the most significant provision is Article 22, which restricts fully automated decisions. If your AI makes decisions that significantly affect individuals (loan denials, job rejections, healthcare triage), you must either obtain explicit consent, demonstrate legal necessity, or provide human involvement. You must also provide meaningful information about the logic involved, which requires explainability.

HIPAA (US Health Insurance Portability and Accountability Act): HIPAA protects health information in the US. If your AI processes protected health information (PHI), you must comply with the HIPAA Security Rule (technical safeguards, encryption, access controls) and the Privacy Rule (limiting use and disclosure, patient rights). Business associate agreements are required with third parties. Violations carry civil and criminal penalties.

CCPA/CPRA (California Consumer Privacy Act / California Privacy Rights Act): California's privacy laws grant consumers rights similar to the GDPR: the right to know what data is collected, the right to delete, the right to opt out of sales, and the right to non-discrimination. CPRA adds restrictions on automated decision-making and profiling. Multi-state privacy laws are emerging with similar provisions.

Cross-border data transfers: If your AI processes data across borders, you must comply with data transfer rules. GDPR restricts transfers outside the EU unless adequate protections are in place (adequacy decisions, standard contractual clauses, binding corporate rules). Transferring EU data to US cloud providers requires careful compliance. China's data localization laws require certain data to remain in China. Map your data flows and ensure compliance with transfer rules.

9.6 Algorithmic Accountability & Transparency Reqs.

Data Protection and Privacy Landscape

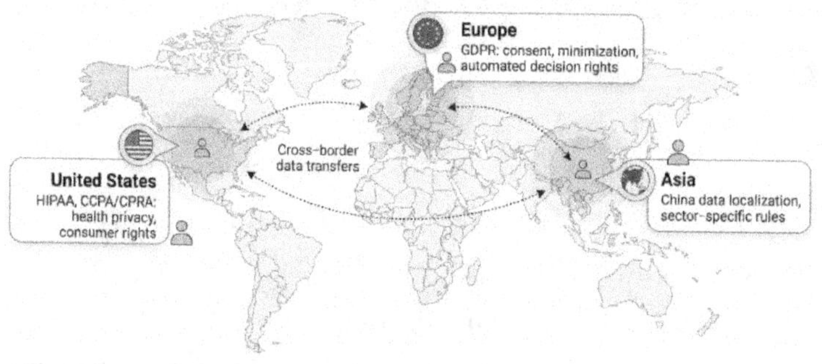

Navigate global data protection requirements for AI systems.

Beyond data protection, emerging laws impose requirements for algorithmic accountability and transparency. These laws require organizations to disclose, explain, and justify how AI systems make decisions.

Explainability requirements: Multiple laws require that AI decisions be explainable. GDPR Article 22 requires "meaningful information about the logic involved" in automated decisions. The EU AI Act requires high-risk systems to provide "appropriate" transparency to enable users to interpret outputs. New York City's bias audit law requires disclosure of how systems evaluate candidates. For IT managers, this means: document how your models work, provide explanations for decisions (feature importance, decision factors), and ensure explanations are accessible to non-technical audiences (individuals affected, auditors, regulators).

Bias audit requirements: Several jurisdictions require bias testing. NYC Local Law 144 mandates annual bias audits for employment tools, measuring selection rates by race and gender. The EU AI Act requires high-risk systems to be tested for bias. Illinois's Artificial Intelligence Video Interview Act requires disclosure of how video interviews are evaluated. Compliance requires: conducting regular fairness testing, documenting testing methodology and results, and making summaries available to stakeholders (job applicants, regulators).

Algorithmic impact assessments: Some laws require impact assessments before deploying high-risk AI. The EU AI Act requires conformity assessments for high-risk systems. Canada's proposed Artificial Intelligence and Data Act (AIDA) requires impact assessments for high-impact systems, evaluating risks to individuals and society. These assessments resemble Data Protection Impact Assessments (DPIAs) under GDPR but focus on algorithmic harms. They require: describing the system and its purpose, identifying

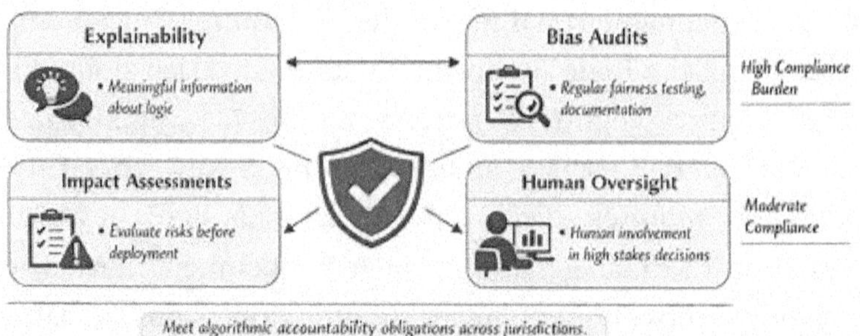

Meet algorithmic **accountability** obligations across jurisdictions.

risks (bias, discrimination, errors), evaluating severity and likelihood, documenting mitigation measures, and obtaining stakeholder input where appropriate.

Human oversight requirements: Many laws mandate human involvement in high-stakes decisions. The EU AI Act requires high-risk systems to have human oversight. The right to contest automated decisions (GDPR, CCPA) implies human review of appeals. Financial regulations require human review for adverse credit decisions. Compliance requires: implementing human-in-the-loop or human-on-the-loop mechanisms, documenting human oversight procedures, and training reviewers to exercise meaningful judgment.

9.7 Building a Compliance-Once Framework

Complying with dozens of overlapping laws across jurisdictions and sectors is not feasible if you build bespoke solutions for each. The solution is a compliance-once framework: design a foundational set of controls that satisfies the strictest requirements globally, then localize as needed for specific jurisdictions and sectors.

Step 1: Identify the strictest requirements. Map all relevant laws and regulations that apply to your AI portfolio. Identify the most stringent requirements in each domain: data protection (GDPR), transparency (EU AI Act), fairness (ECOA, EEOC), human oversight (EU AI Act, GDPR), technical documentation (EU AI Act, FDA), and auditing (NYC bias audits). These become your baseline. If you meet the strictest requirements, you likely meet less strict ones.

Step 2: Design core controls. Build a core compliance framework that satisfies these baseline requirements: robust data governance (data minimization, consent, security), explainability by default (all models provide explanations), fairness testing (bias audits for all high-risk systems), human oversight (HITL for high-stakes decisions), technical documentation (model cards, system documentation), impact assessments (conduct for high-risk use cases), incident response (detect and respond to compliance violations), and audit trails (log all decisions and actions).

Step 3: Implement globally. Deploy these core controls across your entire AI portfolio, regardless of jurisdiction. This creates a high baseline of compliance and simplifies operations—you do not need separate compliance regimes for different systems or regions.

Step 4: Localize for specific requirements. Layer jurisdiction- and sector-specific controls onto the core framework. For example: in the EU, register high-risk systems in the EU database and conduct conformity assessments; in healthcare, obtain FDA clearance and maintain ISO 13485 quality management; in NYC, conduct annual bias audits and publish summaries; and in finance, provide adverse action notices and maintain FCRA-compliant records. Localization is incremental—you are adding specific requirements, not rebuilding the entire compliance program.

Step 5: Maintain and update. Regulations evolve. Monitor regulatory developments in all relevant jurisdictions. When new laws are proposed or enacted, assess their impact on your framework. Update core controls or add localized controls as needed. Assign ownership for regulatory monitoring to legal or compliance teams with support from AI governance.

This approach scales. As you enter new markets or sectors, you assess local requirements and layer them onto your existing framework. You avoid redundant work and maintain consistent, defensible compliance.

9.8 Docs. and Audit Trails: Backbone of Compliance

Regulators do not take your word that systems are compliant. They require evidence: documentation, audit trails, test results, and records. Comprehensive documentation is the backbone of regulatory readiness. Without it, you cannot demonstrate compliance, defend decisions, or respond to audits.

Technical documentation describes how AI systems work. The EU AI Act

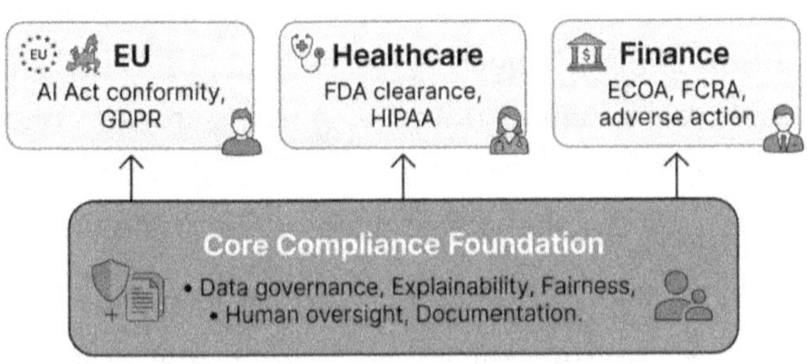

Design global compliance, then adapt to jurisdictions and sectors.

requires detailed technical documentation for high-risk systems, including: system description and purpose, intended users and use cases, data sources and preprocessing methods, model architecture and training procedures, performance metrics and validation results, risk assessment and mitigation measures, human oversight mechanisms, and conformity assessment reports. Maintain this documentation in a centralized repository, update it when systems change, and ensure it is accessible to auditors.

Model cards provide standardized summaries of models: intended use, training data, performance metrics, limitations, fairness evaluations, and ethical considerations. Model cards are not legally required everywhere, but they are best practice and simplify compliance. Create model cards for all deployed models.

Decision logs record individual AI decisions: input data, model output, confidence score, any human review, final decision, and timestamp. Decision logs enable you to audit specific cases, respond to user complaints, investigate incidents, and demonstrate compliance. Implement automated logging for all decisions. Retain logs per regulatory requirements (e.g., GDPR requires logs for the duration of processing plus statute of limitations).

Audit trails track changes to systems: model updates, data changes, configuration changes, policy updates, and approval workflows. Audit trails demonstrate that systems are controlled and that changes follow governance processes. Use version control, model registries, and change management systems to maintain audit trails.

Testing and validation records document that systems were tested and validated before deployment: fairness tests, accuracy evaluations, security assessments, red-team exercises, and user acceptance testing. Retain test reports and results. Regulators may request evidence that systems were validated.

Compliance evidence includes impact assessments, conformity assessments, bias audit reports, DPIA reports, and legal reviews. These documents demonstrate that compliance obligations were fulfilled. Organize them by system and by requirement.

Treat documentation as a first-class deliverable, not an afterthought. Assign ownership for maintaining documentation. Conduct periodic documentation audits to ensure completeness and currency. Good documentation protects you in regulatory investigations and accelerates compliance for new jurisdictions.

Documentation and Audit Trails

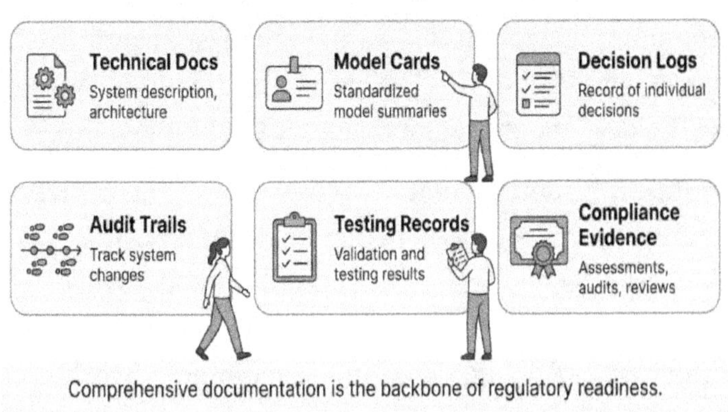

Comprehensive documentation is the backbone of regulatory readiness.

9.9 Responding to Regulatory Audits & Investigations

Even with strong compliance, you may face regulatory audits or investigations. Regulators may conduct routine audits, respond to complaints, or investigate incidents. How you respond determines outcomes. A well-prepared organization can demonstrate compliance and avoid penalties. An unprepared organization faces prolonged investigations and enforcement actions.

Preparation before audits: Maintain compliance documentation current and organized. Conduct internal audits periodically to identify gaps before regulators do. Designate a compliance point of contact who coordinates with regulators. Train teams on how to respond to regulatory inquiries (do not speculate; refer to documentation; coordinate through the compliance lead). Retain legal counsel with regulatory expertise.

Initial response: When you receive an audit notice or investigation inquiry, respond promptly and professionally. Acknowledge receipt, confirm cooperation, and identify the compliance lead who will coordinate. Do not ignore or delay—this signals non-compliance. Review the request carefully: what information is being requested? What is the timeline? What is the scope (specific system, entire portfolio, particular practice)?

Information gathering: Assemble the requested documentation: technical documentation, model cards, decision logs, testing records, policies, and compliance evidence. If you maintained comprehensive documentation (Section 8.7), this should be straightforward. If documentation is incomplete, work urgently to fill gaps. Be transparent about gaps—better to acknowledge them than to be caught concealing information.

Coordination and communication: Coordinate all communication through the designated compliance lead and legal counsel. Do not allow individual engineers or managers to respond to regulators independently—this creates inconsistent messaging and legal risk. Provide clear, factual responses. Avoid defensive or dismissive language. If you do not know an answer, say so and commit to providing it by a specific date.

Demonstrate good faith: Show that you take compliance seriously. Provide evidence of: policies and procedures designed to ensure compliance, training programs for staff, regular testing and audits, incident response and remediation when issues are found, and continuous improvement efforts. Regulators are more lenient with organizations that demonstrate good-faith efforts, even if gaps exist.

Remediation and follow-up: If the audit identifies deficiencies, promptly develop and implement a remediation plan. Provide the plan to regulators with timelines and milestones. Follow through on commitments. Conduct follow-up audits to verify remediation. Document everything. Regulatory audits are learning opportunities—use them to strengthen your compliance program.

9.10   Case: Nav. Multi-Jurisdictional Compliance

To make compliance readiness concrete, consider a case example: a global retail company deploying an AI hiring tool.

Context: The company operates in the US, EU, and Canada. The tool screens resumes, ranks candidates, and recommends interviews. It processes personal data (names, employment history, and education) and makes employment decisions. The company must comply with multiple regulations simultaneously.

Step 1: Identify applicable regulations.

- **EU:** AI Act (high-risk system—employment), GDPR (personal data processing)
- **US:** EEOC guidelines (anti-discrimination), state laws (NYC Local Law 144 requires bias audit)
- **Canada:** Proposed AIDA (high-impact system), PIPEDA (privacy law)

Step 2: Assess requirements.

- **Explainability:** GDPR requires meaningful information, and the AI Act requires transparency
- **Fairness:** EEOC requires non-discrimination, NYC requires bias audit
- **Human oversight:** AI Act requires human oversight, EEOC best practices recommend human review
- **Data protection:** GDPR, PIPEDA require lawful basis, minimization, security
- **Documentation:** AI Act requires technical documentation, NYC requires a bias audit report
- **Impact assessment:** AI Act requires conformity assessment, AIDA requires impact assessment

Step 3: Build a core compliance framework.

- **Data governance:** Collect only necessary data (name, work history, education), obtain consent, implement security controls (encryption, access limits)
- **Explainability:** Model provides feature importance (which resume factors influenced ranking), and candidates receive explanations if not selected
- **Fairness testing:** Conduct bias audits by race and gender, document methodology and results
- **Human oversight:** Implement human-in-the-loop—recruiters review all interview recommendations, can override
- **Technical documentation:** Create model card, system documentation, risk assessment
- **Impact assessment:** Conduct algorithmic impact assessment covering risks, mitigations, stakeholder input

Step 4: Localize for specific jurisdictions.

- **EU:** Register system in EU AI Act database, conduct conformity assessment (internal initially, third-party if required), appoint EU representative if the company is not EU-based
- **NYC:** Conduct annual bias audit, publish summary on website, notify candidates that the tool is used
- **Canada:** Submit an impact assessment to Canadian authorities if AIDA passes

Step 5: Deploy and monitor.

- Deploy system with core compliance controls active globally
- Maintain decision logs for all candidates
- Conduct quarterly fairness monitoring
- Update documentation when system changes
- Monitor regulatory developments in all jurisdictions

Outcome: The company deploys the tool confidently across jurisdictions. Compliance costs are minimized because core controls are built once and localized incrementally. When regulators inquire, comprehensive documentation demonstrates compliance. When NYC law requires annual bias audits, the company is already conducting them. This is regulatory readiness.

9.11   Playbook: Building Regulatory Readiness

You have learned the regulatory landscape, frameworks, and strategies. Now, how do you build regulatory readiness into your organization? Here is a practical playbook.

Step 1: Conduct regulatory mapping. Identify all laws and regulations that apply to your AI portfolio. Consider: jurisdictions where you operate, sectors you serve (healthcare, finance, employment, etc.), data types you process (personal data, health data, biometric data), and AI system characteristics (automated decisions, high-risk, sensitive use cases). Create a regulatory matrix mapping systems to applicable laws.

Step 2: Assess compliance gaps. For each system, evaluate current compliance against regulatory requirements. Use the six-dimensional risk assessment from Chapter 6 as input. Identify gaps: missing documentation, insufficient testing, lack of human oversight, inadequate explainability, and data protection violations. Prioritize gaps by regulatory risk (likelihood of enforcement, severity of penalties).

Step 3: Design compliance-once the framework. Identify the strictest requirements across all applicable laws. Design core controls that satisfy these baseline requirements (data governance, explainability, fairness, human oversight, documentation, impact assessments). Document the framework and communicate it to all AI teams.

Step 4: Implement core controls. Roll out the core compliance framework across your AI portfolio. Update existing systems to meet requirements (may require significant work for legacy systems). Mandate core controls for all new systems (integrate into the approval process from Chapter 2). Assign ownership for each control area.

Step 5: Build localization playbooks. For each jurisdiction and sector with specific requirements, create a localization playbook detailing what additional controls are needed. What documentation or reporting? What timelines? Example: "EU Localization Playbook"—register in database, conduct conformity assessment, appoint representative, maintain EU-specific records. Make playbooks accessible to teams deploying in those contexts.

Step 6: Establish documentation practices. Implement systematic documentation for all systems: technical docs, model cards, decision logs, audit trails, testing records, and compliance evidence. Use templates to standardize. Integrate documentation into workflows (e.g., model card required for model registry submission). Conduct documentation audits quarterly.

Step 7: Monitor regulatory developments. Assign responsibility for tracking regulatory changes (legal, compliance, or AI governance team). Subscribe to regulatory updates, industry groups, and legal newsletters. When new laws are proposed or enacted, assess the impact on your framework. Update controls or add localized requirements as needed. Communicate changes to affected teams.

Step 8: Train and enable teams. Educate AI teams on regulatory requirements: what laws apply? What are their obligations? How to use compliance tools and frameworks? Provide training on data protection, fairness, explainability, and documentation. Make compliance accessible—demystify legal language, provide practical guidance, and offer templates and tools.

Step 9: Build audit readiness. Conduct internal compliance audits annually. Simulate regulatory audits (tabletop exercises). Test whether documentation is complete and accessible. Identify gaps and remediate. Develop response protocols for real audits.

9.12   Common Compliance Pitfalls

Organizations make predictable mistakes in regulatory compliance. Here are five common pitfalls and how to avoid them.

Pitfall 1: Reactive compliance. Waiting until a law is enacted or an audit is announced before addressing compliance. By then, it is too late—remediation is expensive, rushed, and incomplete. **Avoidance:** Build proactive compliance into your AI program from the start. Monitor regulatory trends. Engage early with legal and compliance. Design systems to meet emerging requirements before they are mandatory.

Pitfall 2: Compliance as a checkbox. Treating compliance as a one-time exercise: "We did the DPIA, we are compliant."

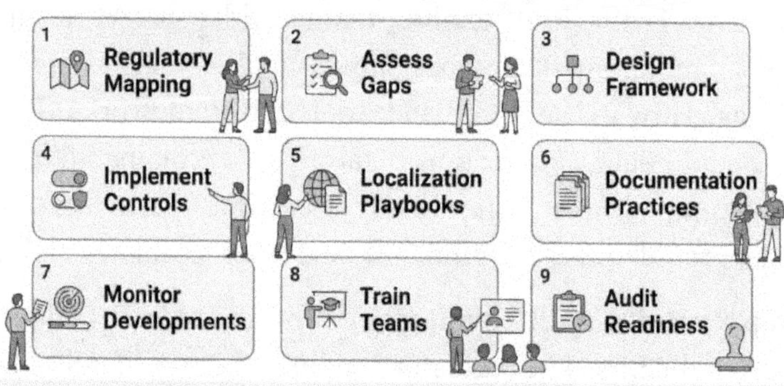

Follow this sequence to build proactive, scalable compliance.

Regulations require ongoing processes: continuous monitoring, regular audits, documentation updates. **Avoidance:** Integrate compliance into operational rhythms. Schedule regular fairness testing, quarterly documentation reviews, annual audits. Make compliance a continuous practice, not a project.

Pitfall 3: Siloed compliance. Different teams (legal, data science, engineering, business) working on compliance independently without coordination, leading to gaps and duplication. **Avoidance:** Create cross-functional compliance teams. Use AI governance structures (Chapter 2) to coordinate. Ensure all stakeholders understand their roles and responsibilities.

Pitfall 4: Inadequate documentation. Failing to document systems, decisions, and processes adequately. When audits come, scrambling to reconstruct documentation after the fact. **Avoidance:** Make documentation a deliverable from day one. Use templates and automation to reduce burden. Assign ownership and accountability for documentation.

Pitfall 5: Ignoring sector-specific rules. Focusing only on horizontal AI laws (EU AI Act, GDPR) and overlooking sector regulations (FDA, ECOA, EEOC). Sector rules often have stricter requirements and more aggressive enforcement. **Avoidance:** Conduct thorough regulatory mapping that includes sector frameworks. Engage sector compliance experts (healthcare compliance, financial compliance) in AI governance.

9.13   AI Reality Check

Before we close, let's address common myths about AI compliance:

- **Myth 1: "Regulations are still years away; we have time."**
  Reality: Major regulations are already in force (GDPR,

HIPAA, ECOA, NYC bias audit law). The EU AI Act's high-risk requirements are phasing in now. Delay creates risk.

- **Myth 2: "Compliance will slow down innovation."**
Reality: Good compliance frameworks enable innovation by providing clear guardrails. Teams know what is acceptable and can move fast within those boundaries. Non-compliance creates uncertainty and last-minute rework, which truly slows innovation.
- **Myth 3: "We are not in the EU, so the AI Act does not apply."**
Reality: If you serve EU customers or process EU residents' data, the AI Act applies regardless of where your company is based. Global companies must comply with global rules.
- **Myth 4: "Our legal team will handle compliance."**
Reality: Compliance is a technical and operational challenge, not just legal. Legal must partner with AI governance, data science, and engineering. IT managers play a central role in implementing compliance controls.

Use this reality check to set expectations and build urgency for regulatory readiness.

9.14   Proactive Readiness for a Regulated Future

This chapter has shown you how to navigate the complex, evolving regulatory landscape for AI. You learned the major regulatory frameworks (EU AI Act, sector regulations, data protection laws, algorithmic accountability requirements), the compliance-once approach for building global, scalable compliance, the critical role of documentation and audit trails, how to respond to regulatory audits, and a practical playbook for building regulatory readiness into your organization.

AI regulation is not a barrier—it is a reality that well-managed organizations turn into a strategic advantage. Proactive compliance builds trust with customers, regulators, and stakeholders. It reduces legal and operational risk. It enables expansion into new markets and sectors with confidence. Reactive organizations will scramble, face enforcement actions, and struggle to operate. Proactive organizations will design compliance once, localize as needed, and scale AI safely across jurisdictions.

As an IT manager, your role is to ensure that regulatory readiness is embedded in your AI program from the start. Compliance is not the legal team's problem or an afterthought—it is a first-class operational requirement. Build it into governance, design, testing, documentation, and monitoring. When regulations evolve or auditors come, you will be ready. That readiness is the difference between AI that operates confidently at scale and AI that becomes a compliance crisis.

10 Embedding Guardrails into the AI Lifecycle

10.1   When Guardrails Are Bolted On, Not Built In

Your data science team just finished training an improved fraud detection model. Accuracy is up 3 percentage points, false positives are down, and the business is eager to deploy. The team submits the model for governance review. That is when the delays start.

Governance requests a fairness audit—no one thought to test for bias during development. Security needs to review the training data provenance—documentation is incomplete. Legal wants an impact assessment—it was never conducted. Compliance asks for explainability testing—the model is a black box. Each review takes two weeks. Meanwhile, fraud losses continue with the old model. The business is frustrated: "Why does governance slow everything down?" The real problem is not governance—it is that guardrails were bolted on at the end instead of built into the development process from the start.

This chapter is about embedding guardrails into the AI lifecycle so they become infrastructure rather than obstacles. You will learn how to integrate testing, validation, and approval processes into CI/CD pipelines, model registries, and data workflows. You will learn about policy as code, continuous validation, and rollout gates that automatically enforce safeguards as systems change. By the end, you will know how to make compliance invisible to developers—guardrails that run automatically, catch issues early, and enable fast, safe deployment. This is where governance meets engineering reality.

10.2   AI Lifecycle: From Experimentation to Prod.

AI systems have a lifecycle distinct from traditional software. Understanding the lifecycle is essential for embedding guardrails at the right points. The AI lifecycle has six major phases: **experimentation, development, validation, deployment, monitoring, and iteration**.

Experimentation is where data scientists explore problems, prototype solutions, and test feasibility. Work is rapid, informal, and exploratory. Models may never reach production. Guardrails should be lightweight but present: data access controls (scientists cannot access production customer data without approval), responsible AI principles education, and sandboxed environments that prevent accidental deployment.

Development is where promising experiments become production candidates. Data scientists prepare production-quality code, document models, and conduct initial testing. Guardrails intensify: code review for security and quality, fairness and bias testing, explainability evaluation, technical documentation (model cards), and version control for reproducibility.

Validation is the checkpoint before production. Models undergo rigorous testing, governance review, and approval. Guardrails are the strictest: comprehensive testing (accuracy, fairness, security, robustness), compliance review (legal, privacy, regulatory), impact assessment (DPIA, algorithmic impact assessment), and governance approval (AI Working Group or Steering Committee sign-off).

Deployment is the transition from development to production. Models are integrated into operational systems, infrastructure is provisioned, and rollout begins. Guardrails ensure safety: deployment gates (validation must pass before deployment is allowed), staged rollout (canary or blue-green deployment), monitoring instrumentation (logging, metrics, alerts), and rollback procedures (automated reversion if issues detected).

Monitoring is continuous observation of production systems. Guardrails track system health, model performance, fairness, and business outcomes (as discussed in Chapter 7). Monitoring generates data that feeds the next phase.

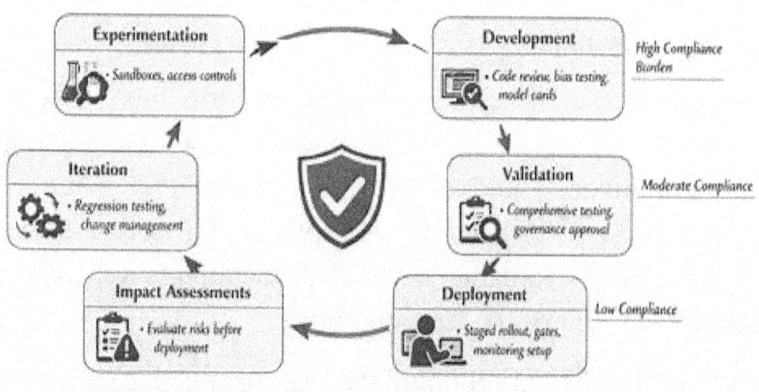

Embed guardrails at every phase for safe, fast AI development.

Iteration is the cycle of improvement: retraining models, updating features, fixing bugs, and responding to drift. Guardrails ensure that changes do not introduce new risks: regression testing (ensure new version does not degrade performance or fairness), change management (document and approve changes), and A/B testing or shadow deployment (validate improvements before full rollout).

For IT managers, the key insight is that guardrails must be embedded at every phase, not just validation. Early-phase guardrails are lightweight and educational. Late-phase guardrails are strict and automated. This balance enables speed without sacrificing safety.

10.3   The Model Registry: Central Hub for Governance

A model registry is a centralized repository where trained models are stored, versioned, documented, and tracked. It serves as the single source of truth for what models exist, where they are deployed, and what their status is. For governance, the model registry is the control point where guardrails are enforced before models reach production.

Core functions of a model registry:

- **Version control:** Track every version of every model. Store model artifacts (weights, parameters, code), training data references, and training configurations. Enable reproducibility—any version can be recreated or rolled back.
- **Metadata management:** Store metadata for each model: owner, use case, training date, performance metrics, fairness scores, approval status, deployment location. Make metadata searchable so governance teams can answer questions like "Which models process customer health data?" or "Which models have not been audited in over a year?"
- **Approval workflow:** Integrate governance approval into the registry. Models cannot be deployed until they pass validation gates and receive governance sign-off. The registry tracks approval history: who approved, when, and based on what evidence.
- **Lineage tracking:** Record the full lineage of each model: what data was used, what preprocessing steps, what hyperparameters, what previous versions influenced it. Lineage enables impact analysis—if training data is found to

be compromised, you can immediately identify all affected models.

- **Access control:** Restrict who can register, approve, or deploy models. Data scientists can register models, but only governance leads can approve them. Only deployment engineers can promote approved models to production.

Integrating guardrails into the model registry:

- **Mandatory documentation:** Models cannot be registered without a completed model card (use case, training data, performance, fairness, limitations).
- **Automated checks:** When a model is registered, automated checks run: Does it meet minimum accuracy thresholds? Does fairness testing show no violations? Is security scanning clean? If checks fail, registration is blocked.
- **Approval gates:** Models move through stages: Registered → Validated → Approved → Deployed. Each transition requires passing criteria and approvals. Governance teams review Validated models and promote them to Approved.
- **Audit trail:** Every action is logged: who registered a model, what tests were run, what results were obtained, who approved, when deployment occurred. This audit trail satisfies regulatory documentation requirements (Chapter 8).

Popular model registry tools include MLflow Model Registry, Azure Machine Learning Registry, AWS SageMaker Model Registry, and open-source solutions like DVC. Choose a tool that integrates with your existing ML infrastructure and supports governance workflows.

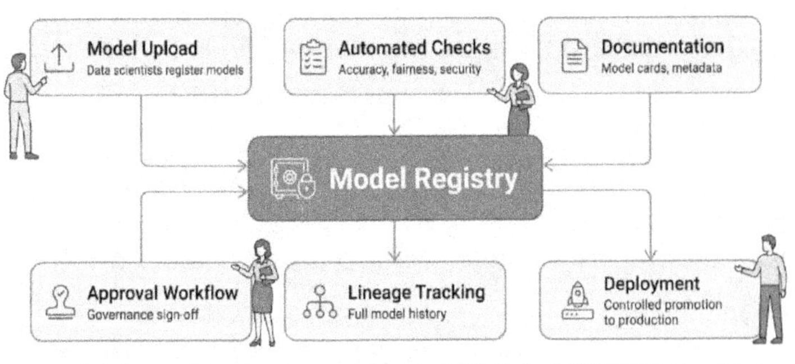

Centralize governance through a model registry with built-in guardrails.

10.4 CI/CD Pipelines for AI: Automation with Safety

Continuous Integration and Continuous Deployment (CI/CD) pipelines automate the process of building, testing, and deploying software. For AI, CI/CD pipelines must be adapted to include model-specific testing and validation. Guardrails are embedded as automated gates: if tests fail, the pipeline stops, preventing unsafe models from reaching production.

Traditional CI/CD vs AI CI/CD:

- **Traditional CI/CD:** Code changes → build → unit tests → integration tests → deploy
- **AI CI/CD:** Code + data + model changes → build → unit tests + model tests + fairness tests + security tests → validation → approval gate → deploy

AI-specific pipeline stages:

Stage 1: Data validation. Before training begins, validate that data meets requirements: schema validation (correct fields, data types), quality checks (no missing values beyond thresholds, no outliers beyond bounds), freshness checks (data is recent enough), and privacy checks (no unauthorized PII, compliance with data governance). If validation fails, stop the pipeline and alert data engineers.

Stage 2: Model training. Train the model using approved training code and validated data. Log training metrics (loss curves, convergence) and hyperparameters. Store trained model artifacts in the model registry.

Stage 3: Model testing. Run comprehensive automated tests:

- **Performance tests:** Accuracy, precision, recall, F1 on holdout data. Compare to baseline. Fail if performance degrades.
- **Fairness tests:** Measure demographic parity, equalized odds, or other fairness metrics across protected groups. Fail if fairness violations exceed thresholds.
- **Robustness tests:** Test on edge cases, adversarial examples, out-of-distribution data. Fail if model behaves unexpectedly.
- **Security tests:** Scan for vulnerabilities, test input validation, check for data leakage. Fail if security issues detected.
- **Explainability tests:** Verify that explanations can be generated for predictions. Fail if explainability requirements are not met.

Stage 4: Approval gate. If all automated tests pass, the model enters a manual approval queue. Governance reviewers examine test results, documentation, and impact assessments. If approved, the model is promoted to deployment-ready status. If rejected, feedback is provided and the model returns to development.

Stage 5: Deployment. Approved models are deployed to production using staged rollout strategies (canary deployment to 5

AI CI/CD Pipeline with Embedded Guardrails

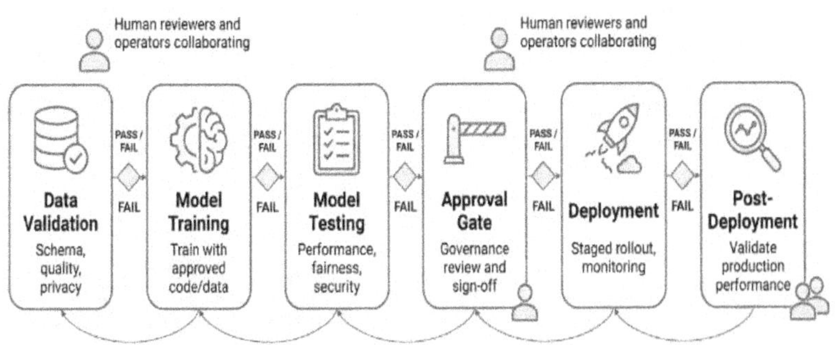

Automate guardrails as pipeline gates to catch issues early.

percent of traffic, monitor for issues, then full rollout). Monitoring instrumentation is activated. Rollback procedures are ready.

Stage 6: Post-deployment validation. After deployment, run shadow tests or A/B comparisons to verify that production performance matches testing. Alert if discrepancies are detected.

Fail-fast principle: Embed guardrails early in the pipeline. If data quality is poor, fail at Stage 1—do not waste resources training a model that cannot be deployed. If fairness tests fail at Stage 3, stop immediately—do not wait for manual review. This saves time and focuses attention on fixable issues.

- 9.4 Policy as Code: Making Guardrails Executable

Policy as code means translating governance policies from documents into executable code that can be automatically enforced. Instead of relying on humans to remember and apply policies,

policies are encoded in automated checks that run every time a model is trained, tested, or deployed. This makes compliance invisible—guardrails enforce themselves.

Examples of policy as code:

- **Fairness policy:** "Models must not exhibit demographic parity violations greater than 5 percentage points between protected groups." → Automated test: calculate selection rates for each group, measure gap, fail if gap > 5 percent.
- **Data minimization policy:** "Models must not use more than 20 features, and no sensitive attributes (race, gender, religion) may be direct inputs." → Automated check: count features, scan for prohibited attributes, fail if violations are detected.
- **Explainability policy:** "All high-risk models must provide feature importance scores for predictions." → Automated test: query model for explanation, fail if explanation cannot be generated.
- **Model refresh policy:** "Models must be retrained at least every 90 days." → Automated alert: check model training date, alert if age > 90 days, block deployment of stale models.

Implementing policy as code:

Step 1: Formalize policies. Review governance policies (Chapter 3) and identify those that can be automated. Break policies into testable conditions: "If X, then Y must be true." For example, "If the model is high-risk, then fairness tests must pass, and human oversight must be present."

Step 2: Write test code. Implement policies as automated tests using testing frameworks (pytest, unittest) or dedicated policy engines (Open Policy Agent, Conftest). Example fairness test:

python

```python
def test_demographic_parity(model, test_data, protected_attribute):
    results = model.predict(test_data)
    group_rates = calculate_selection_rates(results, protected_attribute)
    max_gap = max(group_rates) - min(group_rates)
    assert max_gap <= 0.05, f"Fairness violation: gap={max_gap}"
```

Step 3: Integrate into pipelines. Add policy tests to CI/CD pipelines. Run tests automatically at appropriate stages: data validation tests before training, fairness and security tests after training, deployment policy checks before rollout.

Step 4: Maintain and version policies. Store policy code in version control alongside application code. When policies change, update the test code and re-run pipelines. Use semantic versioning to track policy changes.

Step 5: Monitor policy violations. Log all policy test results. Track violation rates over time. If certain policies fail frequently, investigate: Is the policy too strict? Are teams trained properly? Is the policy unclear? Use metrics to improve policies and training.

Benefits of policy as code: Consistency (policies applied uniformly across all models), speed (automated checks run in seconds), auditability (test results provide compliance evidence), and scalability (policies scale to hundreds of models without additional manual effort).

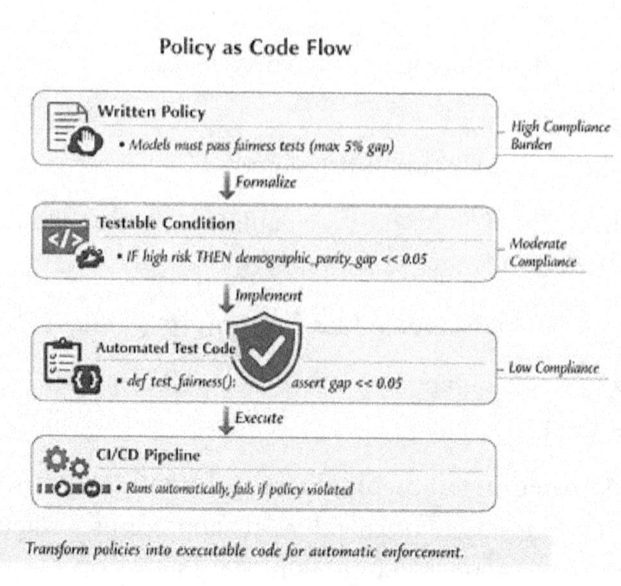

10.5 Data Governance: Controlling the Foundation

Models are only as good as the data they are trained on. Data governance ensures that training data is high-quality, compliant, secure, and ethically sourced. Embedding data governance into the AI lifecycle means catching data issues before they become model issues.

Data governance principles for AI:

- **Data inventory:** Maintain a catalog of all datasets used for AI: what data exists, where it is stored, who owns it, what it contains, and what restrictions apply.
- **Data lineage:** Track data from source to model. If a data source is compromised or found non-compliant, lineage tracking enables rapid impact assessment.
- **Data quality:** Validate data quality continuously. Automated checks for completeness, accuracy, consistency, and timeliness. Poor data quality leads to poor models.
- **Data access controls:** Not all data should be available to all teams. Sensitive data (customer PII, health records) requires approval and logging. Enforce least-privilege access.
- **Data retention and deletion:** Comply with data retention policies and deletion requests (GDPR right to erasure). If data is deleted, models trained on that data may need to be retrained or decommissioned.

Integrating data governance into pipelines:

Pre-training data validation: Before training, validate that the datasets:

- Are approved for AI use (legal and compliance sign-off)
- Meet quality thresholds (no excessive missing values, outliers within bounds)
- Are properly anonymized or de-identified if required
- Do not contain prohibited data (unauthorized PII, embargoed data)
  If validation fails, block training and alert data stewards.

Data versioning: Version datasets just like code. Store snapshots of training data. Record which data version was used for each model. This enables reproducibility and impact analysis.

Sensitive data handling: For sensitive data, use privacy-enhancing techniques such as anonymization, pseudonymization, differential privacy, and synthetic data generation. Enforce encryption in transit and at rest. Log all access for audit.

Data drift detection: Monitor incoming data for drift (Chapter 7). If data distributions change significantly, trigger alerts and consider retraining. Integrate drift detection into data pipelines.

Data governance tools: Use data catalogs (Collibra, Alation, Azure Purview), data quality tools (Great Expectations, Deequ), and data lineage platforms (Apache Atlas, DataHub). Integrate these tools with ML pipelines to automate checks.

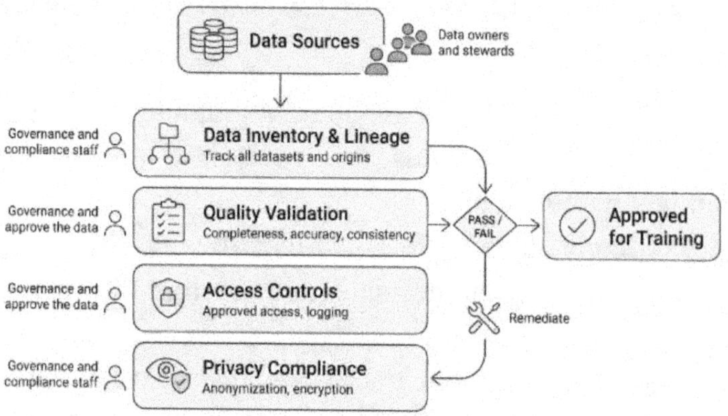

Embed data governance to control the foundation of AI quality.

10.6 Continuous Validation: Testing Beyond Deploy

Traditional software testing ends at deployment. AI requires continuous validation: ongoing testing of production models to ensure they remain accurate, fair, and safe as data and environments evolve. Continuous validation integrates monitoring (Chapter 7) with automated response.

Components of continuous validation:

Automated production testing: Run regular tests on production models: sample predictions and verify correctness (spot-check against ground truth), measure performance metrics on recent data (accuracy, precision, recall), run fairness tests on recent decisions (demographic parity, false positive parity), and check for drift (compare input distributions to training data). Schedule these tests daily, weekly, or continuously, depending on risk.

Canary analysis: When deploying a new model version, run it in parallel with the old version on a small percentage of traffic (canary). Compare performance, fairness, and business metrics between versions. If the new version underperforms, automatically roll back. If it outperforms, gradually increase traffic until full rollout.

Shadow testing: Run new models in shadow mode—process real inputs but do not serve predictions to users. Compare shadow predictions to production predictions. Identify discrepancies and investigate. Shadow testing catches issues before they affect users.

Synthetic testing: Generate synthetic test cases that cover edge cases, adversarial inputs, and known failure modes. Run these tests regularly against production models. Track pass rates over time. Declining pass rates indicate degradation.

User feedback integration: Collect user feedback (complaints, overrides, satisfaction scores) and analyze patterns. If users frequently override or complain about specific types of decisions, investigate. User feedback is a signal that models may be misaligned.

Automated remediation: When continuous validation detects issues, trigger automated responses: alert governance and engineering teams, pause the model and route to human review (circuit breaker), automatically roll back to previous version if degradation is severe, or schedule retraining if drift is detected. Balance automation with human judgment—some issues require investigation, not immediate rollback.

10.7   Rollout Gates and Staged Deployment

Deploying AI changes is risky. New model versions, feature updates, or

Test continuously and respond automatically to maintain quality.

configuration changes can introduce errors, degrade performance, or create unfair outcomes. Rollout gates and staged deployment strategies minimize risk by controlling how changes reach users.

Rollout gates are checkpoints that must be passed before a change can proceed to the next stage. Gates are automated or manual:

- **Automated gates:** Tests that must pass (all unit tests green, fairness tests pass, performance exceeds baseline, security scans clean). If tests fail, deployment is blocked automatically.
- **Manual gates:** Human approvals required (governance sign-off, business owner approval, security review). Until approvals are granted, deployment cannot proceed.

Staged deployment strategies roll out changes gradually, reducing blast radius if issues occur.

Canary deployment: Deploy new version to a small percentage of traffic (5–10 percent). Monitor performance, errors, and user feedback. If metrics are stable or improved, increase traffic incrementally (10 percent $\rightarrow$ 25 percent $\rightarrow$ 50 percent $\rightarrow$ 100 percent). If issues detected, roll back immediately. Canary deployment catches problems with limited user impact.

Blue-green deployment: Maintain two production environments: blue (current version) and green (new version). Deploy new version to green. Route a small percentage of traffic to green for testing. If stable, switch all traffic to green. Blue remains ready for instant rollback if needed. This strategy minimizes downtime and risk.

Feature flags: Use feature flags (toggles) to enable or disable new AI features dynamically without redeploying code. Deploy new model logic behind a flag, initially disabled. Enable the flag for a small user cohort. Monitor. Gradually expand. If issues arise, disable the flag instantly. Feature flags provide fine-grained control and rapid rollback.

Shadow deployment: Deploy new version but do not serve predictions to users. Instead, log predictions for comparison with production version. Analyze differences. Validate that new version behaves as expected. Once validated, promote to canary or full deployment. Shadow deployment provides zero-risk validation.

A/B testing: Randomly assign users to control (old version) or treatment (new version) groups. Compare business metrics (conversion rate, satisfaction) and model metrics (accuracy, fairness). If treatment performs better, roll out fully. If not, revert. A/B testing provides statistically rigorous validation.

Choose strategies based on risk: high-risk changes require canary + shadow + A/B testing. Low-risk changes may use feature flags or blue-green. Always have rollback procedures ready.

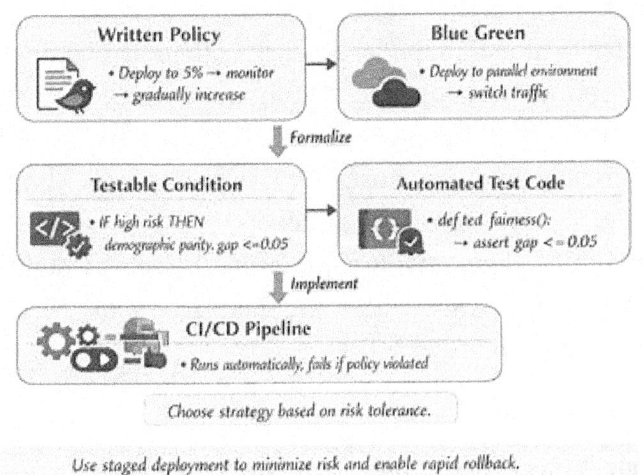

10.8   Case Example: End-to-End Embedded Guardrails

To make these concepts concrete, consider an end-to-end example: a customer service routing system that classifies support tickets and routes them to appropriate teams.

Phase 1: Experimentation

- Data scientists explore ticket data in a sandbox environment with synthetic data (no access to production customer data).
- Prototype classification models, test feasibility.
- Guardrails: Data access controls prevent unauthorized access to customer data. Responsible AI training educates the team on the risks of bias.

Phase 2: Development

- Team prepares production-quality code, conducts initial testing.
- Registers model in model registry with model card: use case description, training data (anonymized customer tickets), performance (90 percent accuracy), fairness evaluation (no significant bias by customer demographics), limitations (struggles with non-English tickets).
- Guardrails: Code review for security. Fairness testing shows demographic parity within acceptable bounds. Model card completed and reviewed.

Phase 3: Validation

- Model enters CI/CD pipeline. Automated tests run:
  - Data validation: Training data passes quality checks (no missing values, schema correct).
  - Model testing: Accuracy meets baseline (90 percent vs 87 percent previous). Fairness tests pass (demographic parity gap 3 percent, below 5 percent threshold). Security scan is clean. Explainability verified (feature importance available).
- Tests pass. Model enters governance approval queue.
- AI Working Group reviews model card, test results, and impact assessment. Approves deployment model.
- Guardrails: Comprehensive automated testing catches issues. Manual governance review ensures alignment with policy.

Phase 4: Deployment

- Model promoted to deployment-ready status in the registry.

- Deployment pipeline uses canary strategy: Deploy to 5 percent of tickets. Monitor routing accuracy, customer satisfaction, and escalation rates.
- Canary metrics are stable after 48 hours. Increase to 25 percent, then 50 percent, then 100 percent over one week.
- Monitoring instrumentation active: Tracks accuracy, fairness, routing times, and user feedback.
- Guardrails: Staged rollout limits risk. Monitoring enables rapid detection of issues. Rollback procedure ready.

Phase 5: Monitoring

- Continuous validation runs daily: Sample 100 tickets, verify routing correctness (95 percent correct), measure demographic parity (gap 2 percent, within bounds), check for data drift (input distributions stable).
- Weekly review of metrics dashboard by the governance team.
- Guardrails: Continuous validation detects degradation early. Regular governance reviews maintain oversight.

Phase 6: Iteration (Month 3)

- Monitoring detects concept drift: Routing accuracy drops to 85 percent due to new ticket types (product launch introduced new categories).
- Alert triggers. Team investigates.
- Decision: Retrain model with recent data including new categories.
- Retraining follows same lifecycle: Development $\rightarrow$ automated tests $\rightarrow$ governance approval $\rightarrow$ canary deployment $\rightarrow$ full rollout.

- Guardrails: Drift detection triggers response. Full lifecycle repeated for retraining ensures safety.

Outcome: Guardrails are invisible to users and mostly invisible to developers. Policies enforce themselves through automated tests. Issues are caught early. Governance reviews are efficient because evidence is pre-assembled. The system operates safely at scale.

10.9   Playbook: Putting Guardrails into Your Lifecycle

You have learned the principles, tools, and strategies. Now, how do you embed guardrails into your organization's AI lifecycle? Here is a practical playbook.

Step 1: Map your current AI lifecycle. Document how your organization currently develops, tests, and deploys AI: What are the phases? What tools are used? Where are manual gates? Where are gaps? Identify pain points: Where do guardrails slow things down? Where do issues slip through?

Step 2: Define guardrail requirements. Based on your governance policies (Chapter 3) and risk assessments (Chapter 6), define the guardrails needed for each lifecycle phase. Example: All high-risk models must pass fairness tests before deployment. Document requirements clearly.

Step 3: Select and implement tools. Choose tools that integrate with your existing infrastructure: Model registry (MLflow, SageMaker, Azure ML), CI/CD platforms (Jenkins, GitLab CI, GitHub Actions), testing frameworks (pytest, Great Expectations), policy engines (Open Policy Agent), and monitoring platforms

(Prometheus, Datadog, custom dashboards). Implement incrementally—start with one use case or team.

Step 4: Build CI/CD pipelines with embedded gates. Design pipelines that include data validation, model training, automated testing (performance, fairness, security), approval gates, and staged deployment. Implement fail-fast: If tests fail, stop immediately. Provide clear feedback to developers on what failed and why.

Step 5: Implement policy as code. Identify policies that can be automated. Write test code for each policy. Integrate tests into pipelines. Version control policy code. Train teams on how policies are enforced.

Step 6: Integrate data governance. Connect data catalogs, quality tools, and access controls to ML pipelines. Validate data before training. Track lineage. Enforce privacy and security requirements.

Step 7: Enable continuous validation. Implement automated production testing, canary analysis, and shadow testing. Define thresholds for acceptable performance and fairness. Automate alerts and rollback procedures.

Step 8: Train teams on new workflows. Developers and data scientists need training on: how to use the model registry, what automated tests will run and how to interpret results, how to write model cards and documentation, what approval gates exist and how to navigate them, and how to use staging and rollback tools. Provide documentation, workshops, and office hours.

Step 9: Monitor and iterate. Track pipeline metrics: How long do pipelines take? How often do tests fail? What are common failure reasons? Use metrics to identify bottlenecks and improve.

Continuously refine policies, tests, and workflows based on feedback.

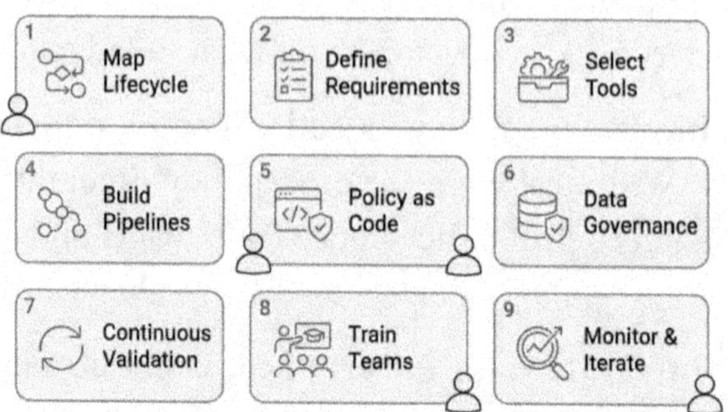

10.10 Common Pitfalls in Lifecycle Integration

Organizations make predictable mistakes when embedding guardrails. Here are five common pitfalls and how to avoid them.

Pitfall 1: Bolting on guardrails at the end and waiting until models are "done" before considering fairness, security, or compliance. By then, fixing issues requires rework, delays deployment, and frustrates teams. **Avoidance:** Embed guardrails from the start. Fairness testing during development, not just before deployment. Security reviews during design, not after code is written. Documentation is built as models are built, not as an afterthought.

Pitfall 2: Over-automating without human judgment and relying entirely on automated tests without manual governance review. Automated tests catch many issues, but not all—some require human judgment (ethical considerations, context-specific risks). **Avoidance:** Combine automated checks with manual gates for high-risk use cases. Use automation to pre-screen and assemble evidence, but retain human decision authority.

Pitfall 3: Fragmented tools and workflows. Different teams use different tools (one team uses MLflow, another uses SageMaker, another uses spreadsheets). Fragmentation makes governance inconsistent and audit trails incomplete. **Avoidance:** Standardize on a core set of tools. Integrate them into a unified workflow. Provide training and support to encourage adoption.

Pitfall 4: Ignoring data governance and focusing on guardrails on models while neglecting data quality, lineage, and compliance. Poor data leads to poor models, regardless of model-level guardrails. **Avoidance:** Treat data governance as foundational. Validate data before training. Track lineage. Enforce access controls and privacy requirements.

Pitfall 5: Static guardrails that never evolve. Implementing guardrails once and never updating them as policies, risks, or technologies change. Static guardrails become outdated and ineffective. **Avoidance:** Treat guardrails as living infrastructure. Review and update policies quarterly. Incorporate lessons learned from incidents. Continuously refine tests and thresholds in response to feedback.

10.11 AI Reality Check

Before we close, let's address common myths about embedding guardrails:

- **Myth 1: "Automation means no human involvement."**
  Reality: Automation handles routine checks, but humans remain essential for judgment, approval, and oversight. Automation enables humans to focus on high-value decisions.

- **Myth 2: "Guardrails slow down development."**
  Reality: Poorly designed guardrails (manual, late-stage) slow things down. Well-designed guardrails (automated, early-stage) actually speed development by catching issues early, when they are cheap to fix.

- **Myth 3: "We need perfect policies before we can automate."**
  Reality: Start with imperfect policies and iterate. Implement basic automated checks (accuracy thresholds, basic fairness tests), learn from results, and refine. Perfection is the enemy of progress.

- **Myth 4: "Embedding guardrails is a one-time project."**
  Reality: Embedding guardrails is an ongoing capability. Policies evolve, tools improve, and new risks emerge. Treat it as continuous improvement, not a finite project.

Use this reality check to set expectations and build commitment to lifecycle integration.

10.12 Summary: Guardrails as Infrastructure

This chapter has shown you how to embed guardrails into the AI lifecycle so they become infrastructure rather than obstacles. You learned the six phases of the AI lifecycle (experimentation, development, validation, deployment, monitoring, iteration), the role of the model registry as a governance hub, how to build CI/CD pipelines with embedded gates, policy as code for automatic enforcement, data governance integration, continuous validation for ongoing safety, and staged deployment strategies to minimize risk.

The key insight is that guardrails are most effective when they are invisible—automated, integrated, and enforced by infrastructure rather than manual processes. When developers push code, tests run automatically. When models are registered, fairness checks are executed. When deployment happens, staged rollout protects users. Guardrails do not slow teams down; they provide confidence to move fast because risks are managed systematically.

As an IT manager, your role is to build and maintain this infrastructure. Invest in tools, pipelines, and automation. Train teams on new workflows. Monitor effectiveness and continuously improve. When guardrails are embedded, compliance becomes a byproduct of good engineering rather than a separate burden. That is when AI governance scales—and when your organization can innovate confidently, knowing that every change is tested, validated, and safe.

If your organization is in Stage 3 (Defined), your goal is to reach Stage 4 (Embedded). This transition requires making guardrails invisible through automation and culture change. It is the hardest transition because it requires deep organizational change, not just process implementation. Here is how to do it.

Step 1: Maximize automation. Push guardrails into infrastructure: implement policy as code (Chapter 9)—all policies enforceable through automated tests, achieve full CI/CD integration—every model goes through automated testing and approval gates, establish continuous validation—production systems tested automatically and continuously (Chapter 7), and automate monitoring and alerting—drift, performance, and fairness issues detected and escalated automatically. The goal is that compliance happens without human intervention for routine cases. Humans focus on judgment calls, exceptions, and improvement.

Step 2: Embed guardrails into culture. Make responsible AI part of how the organization defines itself: include responsible AI in mission and values statements, make responsible AI a topic in onboarding for all employees (not just AI teams), incorporate responsible AI into product development rituals (design reviews, sprint retrospectives), empower teams to stop work if guardrails are threatened (anyone can pull the "stop" lever), and make responsible AI a dimension of leadership competence (leaders are evaluated on how they steward AI). Culture change is slow—it requires consistent reinforcement over years.

Step 3: Create centers of excellence. Establish specialized capabilities that support the organization: a responsible AI research group (stays current on techniques, tools, and regulations), a model risk management team (deep expertise in fairness, explainability, adversarial robustness), and a compliance and regulatory affairs team (interprets laws, guides compliance strategy). Centers of excellence provide advanced support, thought leadership, and continuous innovation.

Step 4: Build continuous learning loops. Make learning systematic: conduct quarterly governance retrospectives (what went well, what did not, what to change), analyze all incidents and near-misses (root cause, lessons learned, preventive actions), track external developments (regulatory changes, peer incidents, research advances), and update policies, processes, and tools based on learnings. Document all changes and communicate them organization-wide. Continuous learning prevents stagnation.

Step 5: Demonstrate thought leadership. Position your organization as a leader in responsible AI: publish thought leadership (blog posts, white papers, conference talks), contribute to industry standards and frameworks, participate in regulatory consultations, and share learnings openly (transparently discuss incidents and how you addressed them). Thought leadership attracts talent, builds trust, and influences the broader ecosystem. It also reinforces internal culture—teams take pride in being leaders.

Step 6: Measure outcomes, not just outputs. Shift from measuring activities (risk assessments completed, training sessions held) to measuring outcomes (incidents prevented, models, decommissioned before harm, customer trust improved, regulatory readiness demonstrated). Use outcome metrics to validate that guardrails are effective, not just present. If metrics show issues (e.g., incidents increasing despite mature processes), investigate and adjust.

The transition to Stage 4 (Embedded) takes 18–24 months of sustained effort after reaching Stage 3. It requires patience, persistence, and a commitment to leadership. But the result is an organization where responsible AI is not a program—it is identity.

11 Culture, Maturity, and Continuous Improvement

11.1   When Guardrails Become Cultural DNA

Two years ago, your organization launched its AI governance program. You built the frameworks, wrote the policies, trained the teams, and implemented the controls. Initially, it felt like pushing a boulder uphill. Data scientists complained about approval delays. Business units resisted fairness testing. Legal questioned every decision. Today, something has shifted. A junior data scientist just rejected her own model because fairness tests showed a 6 percent demographic gap—above the 5 percent threshold—without anyone telling her to. A product manager proactively requested a human-in-the-loop review for a new feature, citing risk considerations. An engineer spotted a production drift and immediately triggered the incident response playbook. No one had to remind them. No one had to enforce compliance. The guardrails have become invisible because they are now cultural DNA. Teams do not see responsible AI as a burden—they see it as a matter of professionalism. This is what maturity looks like.

This chapter is about the journey from bolted-on controls to embedded culture. You will learn the four stages of AI governance maturity—ad hoc, emerging, defined, and embedded—and how organizations progress through them. You will learn how to build the cultural foundations that sustain guardrails: leadership commitment, team enablement, incentive alignment, and continuous learning. You will learn practical strategies for advancing maturity, measuring progress, and maintaining momentum. By the end, you will understand that guardrails are not a project with an end date—

they are an ongoing practice that evolves into organizational identity. The goal is not just to have guardrails, but to become an organization where responsible AI is second nature.

11.2   Why Culture Matters More Than Controls

You can have perfect policies, sophisticated tools, and comprehensive processes, but if your culture does not support them, they will fail. Culture is the invisible force that determines whether guardrails are followed, ignored, or actively resisted. Culture is what people do when no one is watching. It is the assumptions they make, the shortcuts they take, the priorities they set, and the behaviors they reward.

There are three reasons why culture matters more than controls. First, **policies cannot cover every scenario**. AI systems are complex and dynamic. Edge cases, novel situations, and judgment calls emerge constantly. When policies are silent, culture fills the gap. If the culture prioritizes speed over safety, people will skip optional checks. If the culture prioritizes responsibility, people will err on the side of caution.

Second, **controls are only as strong as the people who execute them**. You can mandate fairness testing, but if data scientists do not understand why it matters or do not believe in it, they will treat it as a checkbox—run the test, move on, ignore the results. Engaged teams who understand the mission and internalize the values will execute controls with care and thoughtfulness.

Third, **culture determines resilience**. Organizations face pressure: tight deadlines, budget constraints, competitive urgency, and leadership demands. When pressure mounts, culture determines whether guardrails are sacrificed first or last. A strong culture treats guardrails as non-negotiable. A weak culture treats them as optional.

For IT managers, this means that building guardrails is not just a technical or procedural challenge—it is a cultural challenge. You must invest in mindset, education, incentives, and norms. You must model the behaviors you want to see. You must celebrate teams who prioritize responsibility, even when it slows them down. Culture is built through hundreds of small decisions, reinforced over time, until it becomes the way things are done.

11.3   The Four Stages of AI Governance Maturity

As AI governance develops, organizations progress through a series of identifiable stages. Knowing these phases enables you to evaluate your current position, pinpoint any shortcomings, and chart a course ahead. These four stages are: **ad hoc, rising, defined, and embedded**.

Ad hoc (Stage 1): At this stage, governance is handled reactively and unevenly. Although some policies might be written down, they're rarely enforced systematically. Teams work separately with minimal coordination, and guardrails are only put in place occasionally—usually after issues or audits arise. There's no formal governance framework, standard processes, or clear accountability. Each team takes its own approach, and documentation is sparse or missing altogether. While leadership may recognize the importance of governance, they haven't dedicated resources to support it. This approach carries considerable risks: incidents are probable, compliance remains questionable, and the organization could easily suffer harm.

Rising (Stage 2): The organization recognizes the need for governance and begins building capability. Policies are drafted and communicated. A governance committee is formed. Initial processes are designed (approval workflows, risk assessments). Some teams begin using guardrails, but adoption is uneven. Tools and automation are limited—most guardrails are manual. Documentation improves but remains inconsistent. Leadership provides verbal support but may not yet allocate sufficient resources. This stage is characterized by experimentation and learning: the organization is figuring out what works and what does not. Progress is visible, but fragility remains—if key champions leave or priorities shift, momentum can stall.

Defined (Stage 3): Governance is now formal and consistently applied. Policies are not only documented and authorized but also shared throughout the organization. Processes have been standardized: all AI use cases follow identical approval workflows, use a unified risk assessment framework, and comply with uniform testing requirements. Roles and responsibilities are well-defined. Tools such as model registries, CI/CD pipelines, and monitoring dashboards, along with automation, are established. Training ensures teams know the policies and how to implement them. Documentation is thorough and prepared for audits. Leadership commits resources—a governance team, budget, and executive sponsorship—specifically for governance. Guardrails are integrated into daily workflows, not optional, though compliance still needs deliberate attention. Teams must actively adhere to processes; guardrails are apparent and can sometimes feel restrictive.

Embedded (Stage 4): Governance is seamlessly integrated into the organizational culture and operational processes. Guardrails operate automatically—policies are encoded as code, testing occurs within pipelines, and monitoring is continuous. Teams consistently apply responsible AI principles, making ethical decisions instinctively rather than merely adhering to policy requirements. Leadership regards responsible AI as both a strategic differentiator and an operational standard. Guardrails are viewed not as compliance burdens but as drivers of trust and efficiency. Documentation, testing, and monitoring are natural outcomes of routine activities. The organization fosters ongoing learning: although incidents are infrequent, any that arise are systematically analyzed and addressed through swift improvements. At this stage, responsible AI becomes an essential component of the organization's identity.

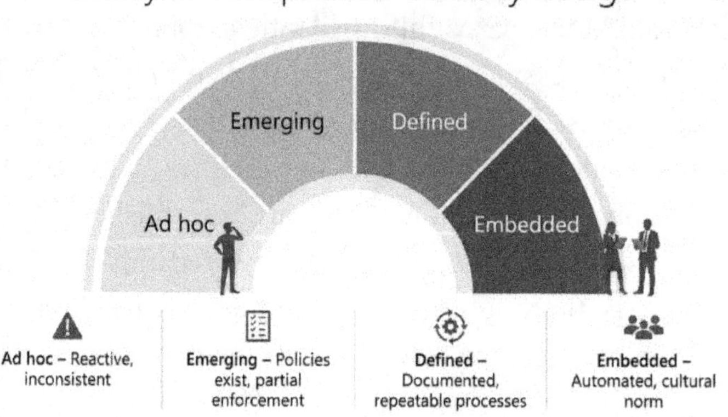

11.4   Assessing Your Current Maturity

Before you can advance maturity, you must understand where you stand. Maturity assessment is a structured evaluation of your current AI governance capabilities across multiple dimensions.

Conduct this assessment annually to track progress and identify improvement priorities.

Assessment dimensions: Evaluate maturity in eight key areas:

Policy and Standards: Are policies documented, communicated, and updated regularly? Are they comprehensive or minimal?

Governance Structure: Is there a formal governance committee with clear roles, decision authority, and accountability?

Processes and Workflows: Are approval, testing, and deployment processes standardized and repeatable? Are they documented?

Tools and Automation: What tools are in place (model registries, pipelines, monitoring)? How much is automated vs manual?

Risk Management: Is risk assessment systematic? Are high-risk use cases treated differently from low-risk?

Compliance and Documentation: Is documentation comprehensive and audit-ready? Are regulatory requirements understood and met?

1. **Training and Culture:** Are teams trained on responsible AI? Do they internalize principles or treat them as checkboxes?

Continuous Improvement: Does the organization learn from incidents? Are processes refined based on feedback?

Assessment method: For each dimension, rate your organization on a 1–4 scale corresponding to the four maturity stages. Use specific evidence:

Stage 1 (Ad hoc): Minimal or no capability. Activities are reactive and inconsistent.

Stage 2 (Emerging): Capability exists but is incomplete or unevenly applied. Initial processes and tools are in place.

Stage 3 (Defined): Capability is formalized, documented, and consistently applied—repeatable processes and comprehensive tools.

Stage 4 (Embedded): Capability is fully integrated, automated, and cultural—continuous improvement and proactive management.

Scoring example:

Policy and Standards: Stage 3 (comprehensive policies documented and communicated, but not yet fully automated)

Governance Structure: Stage 3 (formal committee, clear roles, regular meetings)

Processes and Workflows: Stage 2 (some standardization, but inconsistent execution across teams)

Tools and Automation: Stage 2 (model registry exists, but CI/CD integration is incomplete)

Risk Management: Stage 3 (systematic risk assessment, differentiated controls)

Compliance: Stage 2 (improving documentation, but gaps remain)

Training and Culture: Stage 2 (training programs exist, but adoption is uneven)

Continuous Improvement: Stage 2 (some learning from incidents, but not systematic)

Overall maturity: Calculate the average score. In this example, the average is 2.4—solidly in Stage 2 (Emerging) with some Stage 3 capabilities. This organization is making progress but has work to do to reach Stage 3 (Defined).

Use the assessment to prioritize: Identify the lowest-scoring dimensions. These are your improvement priorities. Focus resources on lifting these areas to match your higher-scoring dimensions.

Assess eight dimensions to identify improvement priorities.

11.5   Building the Cultural Foundations

Advancing maturity requires building cultural foundations that support and sustain guardrails. There are five foundational elements: **leadership commitment, psychological safety, incentive alignment, education and enablement, and visible recognition.**

Leadership commitment means that executives and senior managers demonstrate, through actions and resource allocation, that responsible AI is a priority. Commitment is not just verbal support—it is budget for governance teams, time allocated for training, willingness to delay launches to address guardrail issues, and consistent messaging that safety and ethics are non-negotiable. Leaders must model the behaviors they want to see: attend governance meetings, ask about fairness and risk in reviews, celebrate teams that catch issues early, and hold teams accountable when guardrails are bypassed. Leadership commitment cascades: when executives prioritize responsible AI, managers do the same, and teams follow.

Psychological safety means that people feel safe raising concerns, reporting issues, and admitting mistakes without fear of punishment. AI governance depends on transparency: teams must be willing to surface problems, flag risks, and escalate incidents. If the culture punishes messengers or treats mistakes as career-limiting, people will hide issues until they become crises. Create psychological safety by: responding to bad news with curiosity, not blame ("What happened? What can we learn?"), rewarding teams for catching and reporting issues early, conducting blameless post-incident reviews that focus on systems and processes, and making it clear that responsible AI is everyone's responsibility, not just governance's.

Incentive alignment means that performance management, rewards, and recognition reinforce responsible AI behaviors. If teams are measured solely on speed (time to deploy) and impact (revenue, users), they will optimize for those metrics at the expense of guardrails. Align incentives by: including responsible AI metrics in performance reviews (fairness testing completed, incidents prevented, documentation quality), recognizing teams who demonstrate exceptional responsibility (even when it slows them down), tying bonuses or promotions to governance compliance, and making responsible AI a qualification for leadership roles. What gets measured and rewarded gets done.

Education and enablement mean that teams have the knowledge, skills, and tools to execute guardrails effectively. Training cannot be a one-time onboarding session—it must be ongoing, practical, and role-specific. Data scientists need training on fairness metrics and explainability techniques. Engineers need training on secure coding and privacy. Product managers need training on risk assessment and ethical considerations. Provide: workshops and hands-on labs, documentation and quick-reference guides, office hours with governance experts, case studies and real-world examples, and certification programs for advanced skills. Make learning accessible and applied, not abstract.

Visible recognition means publicly celebrating responsible AI behaviors. Recognition reinforces culture by signaling what the organization values. Create recognition programs: awards for best responsible AI practices, spotlight features in internal communications highlighting teams that exemplify guardrails, executive shout-outs in all-hands meetings for teams that catch issues early, and "guardrail champion" designations for individuals who go above and beyond. Make responsible AI something people are proud of, not something they endure.

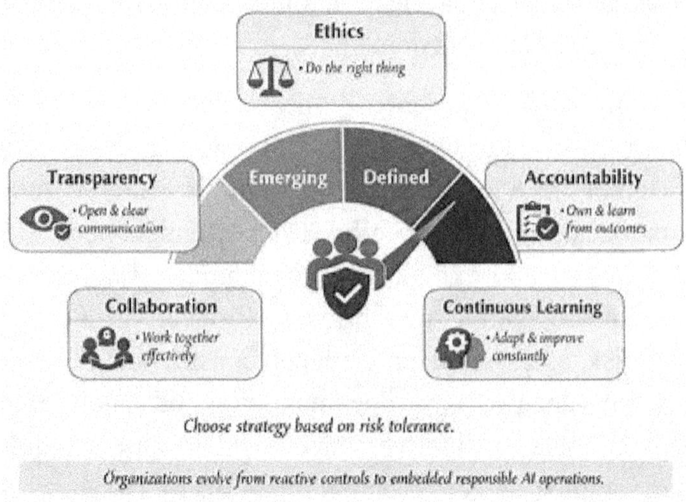

Advancing from Ad Hoc to Emerging

If your organization is in Stage 1 (Ad hoc), your first goal is to reach Stage 2 (Emerging). This transition requires establishing basic structures and demonstrating value. Here is how to do it.

Step 1: Secure executive sponsorship. You cannot advance without leadership support. Make the case to executives: show the risks of unmanaged AI (incidents from peers, regulatory enforcement examples, potential organizational harms), present a roadmap for building governance capability, request modest initial resources (governance lead, budget for tools and training), and demonstrate quick wins (conduct risk assessment for one high-visibility use case, catch and prevent one issue, show cost savings or risk reduction). Secure a sponsor—a senior leader who will champion the effort and provide air cover.

Step 2: Form a governance working group. Assemble a cross-functional team: AI governance lead (chair), data science representation, legal and compliance representation, IT/engineering representation, and business owner representation. Meet bi-weekly initially. Establish charter: define scope (what AI use cases are covered), define decision authority (what this group approves), and define initial goals (e.g., draft policies, conduct risk assessments, pilot guardrails with one team).

Step 3: Draft foundational policies. Create 3–5 core policies that address the highest risks: responsible AI principles (fairness, transparency, accountability), AI use case approval process, data governance for AI (access, quality, privacy), and high-risk AI requirements (human oversight, testing, documentation). Keep policies concise and actionable—2–3 pages each. Get legal review and executive approval.

Step 4: Pilot with a friendly team. Choose one AI use case and one willing team to pilot guardrails. Work with them to: conduct risk assessment, implement basic testing (accuracy, fairness, security), create documentation (model card, use case summary), and obtain governance approval. Use this pilot to refine processes, identify pain points, and demonstrate feasibility. Showcase success to build momentum.

Step 5: Communicate and build awareness. Share what you are doing: publish policies on the intranet, present at team meetings and town halls, create simple guides ("What AI teams need to know about governance"), and invite feedback and questions. Make governance visible and approachable, not intimidating.

Step 6: Track progress and iterate. Measure: number of use cases assessed, policies published and approved, teams trained, and incidents prevented. Review progress monthly with the governance working group. Adjust based on feedback. Celebrate small wins publicly. After 6–12 months of consistent execution, you will have transitioned to Stage 2 (Emerging).

11.6   Advancing from Emerging to Defined

If your organization is in Stage 2 (Emerging), your goal is to reach Stage 3 (Defined). This transition requires scaling and standardizing: moving from experimentation to repeatable processes that apply organization-wide. Here is how to do it.

Step 1: Formalize governance structure. Elevate the governance working group to a formal AI Governance Committee with: a clear charter and decision authority (approved by executives), regular meeting cadence (monthly or bi-monthly), documented roles and responsibilities, and integration with existing governance bodies (IT governance, risk committee, compliance). Appoint dedicated staff (AI governance manager, compliance analyst).

Step 2: Standardize processes. Document and standardize all governance processes: AI use case proposal and approval workflow, risk assessment framework (Chapter 6), testing requirements (performance, fairness, security, explainability), deployment approval process, and monitoring and incident response (Chapter 7). Publish process documentation. Train all teams on processes. Make processes mandatory for all new AI use cases.

Step 3: Implement tools and automation. Deploy tools that scale governance: model registry with approval workflows (Chapter 9), CI/CD pipelines with automated testing (Chapter 9), monitoring dashboards (Chapter 7), and a documentation repository (centralized storage for model cards, assessments, and approvals). Integrate tools into workflows so guardrails become automatic, not manual.

Step 4: Build training programs. Develop role-specific training: data scientists: fairness, explainability, model documentation; engineers: secure deployment, privacy, monitoring; product managers: risk assessment, ethical considerations; executives: responsible AI strategy, regulatory landscape. Deliver training through workshops, e-learning, and certification programs. Track training completion. Make training mandatory for anyone working on AI.

Step 5: Expand to all use cases. Apply governance processes to the entire AI portfolio: conduct portfolio inventory (identify all existing AI use cases), conduct risk assessments for all use cases (using Chapter 6 framework), bring non-compliant use cases into compliance (remediation plans, timelines), and mandate governance for all new use cases (no AI can deploy without approval). This is a significant effort—prioritize high-risk use cases first.

Step 6: Establish metrics and reporting. Define governance metrics: percentage of use cases with completed risk assessments, percentage of high-risk use cases with required testing, number of governance approvals per month, number of incidents detected and resolved, and training completion rates. Report metrics monthly to executive leadership. Use metrics to identify gaps and drive improvement.

After 12–18 months of consistent execution, you will have transitioned to Stage 3 (Defined). Governance is now systematic, repeatable, and organization-wide.

11.7 Advancing from Defined to Embedded

If your organization is in Stage 3 (Defined), your goal is to reach Stage 4 (Embedded). This transition requires making guardrails invisible through automation and culture change. It is the hardest transition because it requires deep organizational change, not just

Maturity Progression Roadmap

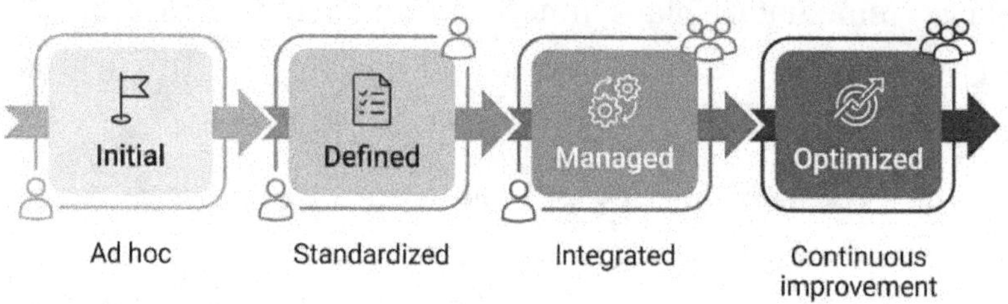

Use staged milestones to plan and communicate maturity progression.

process implementation. Here is how to do it.

Step 1: Maximize automation. Push guardrails into infrastructure: implement policy as code (Chapter 9)—all policies enforceable through automated tests, achieve full CI/CD integration—every model goes through automated testing and approval gates, establish continuous validation—production systems tested automatically and continuously (Chapter 7), and automate monitoring and alerting—drift, performance, and fairness issues detected and escalated automatically. The goal is to ensure compliance without human intervention for routine cases. Humans focus on judgment calls, exceptions, and improvement.

Step 2: Embed guardrails into culture. Make responsible AI part of how the organization defines itself: include responsible AI in mission and values statements, make responsible AI a topic in onboarding for all employees (not just AI teams), incorporate responsible AI into product development rituals (design reviews, sprint retrospectives), empower teams to stop work if guardrails are threatened (anyone can pull the "stop" lever), and make responsible AI a dimension of leadership competence (leaders are evaluated on how they steward AI). Culture change is slow—it requires consistent reinforcement over years.

Step 3: Create centers of excellence. Establish specialized capabilities that support the organization: responsible AI research group (stays current on techniques, tools, and regulations), model risk management team (deep expertise in fairness, explainability, adversarial robustness), and compliance and regulatory affairs team (interprets laws, guides compliance strategy). Centers of excellence provide advanced support, thought leadership, and continuous innovation.

Step 4: Build continuous learning loops. Make learning systematic: conduct quarterly governance retrospectives (what went well, what did not, what to change), analyze all incidents and near-misses (root cause, lessons learned, preventive actions), track external developments (regulatory changes, peer incidents, research advances), and update policies, processes, and tools based on learnings. Document all changes and communicate them organization-wide. Continuous learning prevents stagnation.

Step 5: Demonstrate thought leadership. Position your organization as a leader in responsible AI: publish thought leadership (blog posts, white papers, conference talks), contribute to industry standards and frameworks, participate in regulatory consultations, and share learnings openly (transparently discuss incidents and how you addressed them). Thought leadership attracts talent, builds trust, and influences the broader ecosystem. It also reinforces internal culture—teams take pride in being leaders.

Step 6: Measure outcomes, not just outputs. Shift from measuring activities (risk assessments completed, training sessions held) to measuring outcomes (incidents prevented, models decommissioned before harm, customer trust improved, regulatory readiness demonstrated). Use outcome metrics to validate that guardrails are effective, not just present. If metrics show issues (e.g., incidents increasing despite mature processes), investigate and adjust.

The transition to Stage 4 (Embedded) takes 18–24 months of sustained effort after reaching Stage 3. It requires patience, persistence, and leadership commitment. But the result is an organization where responsible AI is not a program—it is identity.

11.8   Maintaining Momentum and Avoiding Backslide

Maturity is not permanent. Organizations can backslide if attention wanes, leadership changes, or priorities shift. Maintaining momentum requires deliberate effort. Here are strategies to prevent backslide.

Institutionalize governance. Embed governance into organizational structure so it survives personnel changes: secure permanent budget for governance teams (not project funding), codify governance in corporate policies and bylaws, integrate governance into strategic planning and budgeting cycles, and tie governance metrics to executive scorecards. Institutionalization makes governance resistant to deprioritization.

Refresh leadership commitment regularly. Leadership support can erode over time, especially if no visible incidents occur ("We have not had problems, so why do we need this?"). Refresh commitment by: providing quarterly briefings to executives on governance value (risks mitigated, incidents prevented), showcasing external regulatory developments that validate governance investment, highlighting peer incidents that could have happened to you, and bringing in external experts (auditors, regulators, advisors) to validate your approach. Keep governance visible at the leadership level.

Rotate governance roles. Avoid single points of failure. Governance should not depend on one champion. Rotate responsibilities: have multiple people trained on governance processes, rotate governance committee membership every 2–3 years, create succession plans for governance leads, and document processes so knowledge is not tribal. Rotation builds resilience and spreads ownership.

Monitor for warning signs of backslide. Watch for signals that maturity is declining: governance meetings skipped or canceled, guardrails bypassed with increasing frequency (ask for metrics: override rates, approval times), training completion rates dropping, documentation quality declining, or incidents increasing. If warning signs appear, investigate immediately. Diagnose: Is this a capacity issue (too much work, not enough resources)? A culture issue (teams deprioritizing guardrails)? A leadership issue (insufficient support)? Address root causes proactively.

Celebrate persistence, not just results. Responsible AI is a long game. Celebrate teams who

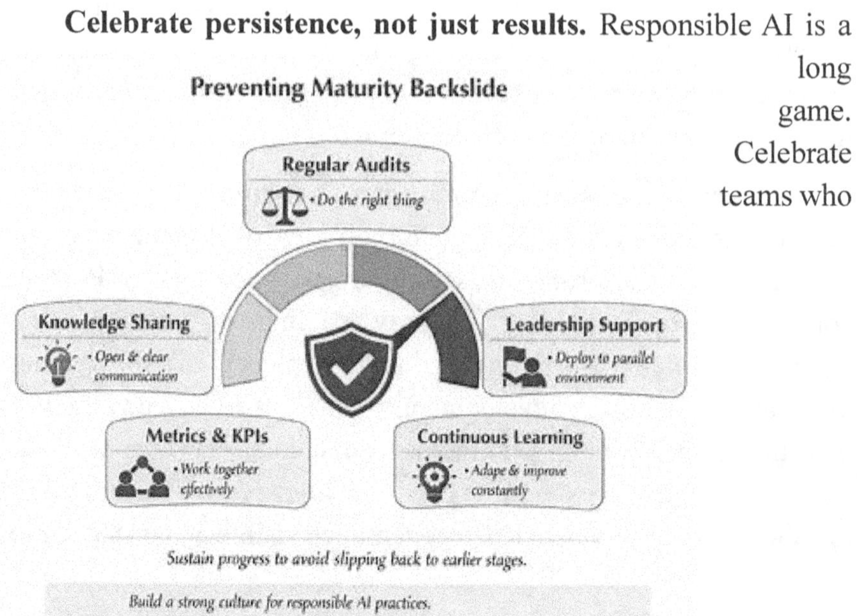

consistently execute guardrails, even when results are not flashy. Recognize the data scientist who caught drift early and triggered retraining before users noticed. Recognize the engineer who flagged a privacy risk during design and prevented a violation. Recognize the manager who delayed a launch to address a fairness issue. Persistence in doing the right thing, day after day, is what sustains maturity.

11.9   Measuring Success: Metrics for Maturity

Measuring governance maturity requires both leading indicators (activities and capabilities) and lagging indicators (outcomes and impact). Track both to understand progress and effectiveness.

Leading indicators (capability):

Policy coverage: Percentage of AI use cases covered by governance policies.

Process compliance: Percentage of use cases that complete required approvals, risk assessments, and testing.

Training completion: Percentage of AI practitioners who have completed responsible AI training.

Tool adoption: Percentage of models registered in model registry, deployed through CI/CD pipelines.

Documentation quality: Percentage of models with complete model cards and technical documentation.

Lagging indicators (outcome):

Incidents prevented: Number of issues caught by guardrails before reaching production (failed tests, governance rejections, monitoring alerts that prevented harm).

Incidents occurred: Number of production incidents related to AI (bias discovered, security breach, compliance violation). Lower is better.

Time to detect: Average time from incident occurrence to detection. Shorter is better (indicates effective monitoring).

Time to resolve: Average time from detection to resolution. Shorter is better (indicates effective response).

Regulatory readiness: Ability to respond to regulatory inquiries or audits without scrambling. Measured by audit preparedness assessments.

Stakeholder trust: User trust, regulator confidence, executive satisfaction with governance. Measured through surveys or feedback.

Balanced scorecard approach: Track 3–5 leading indicators and 3–5 lagging indicators. Report monthly to governance

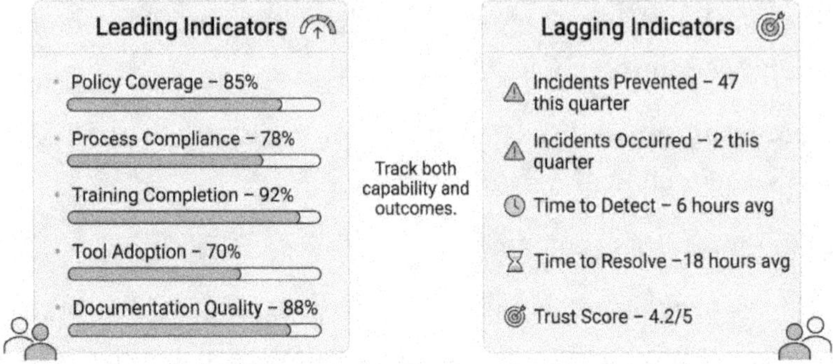

Use balanced metrics to measure governance effectiveness.

committee, quarterly to executives. Use trends, not point-in-time values: Are indicators improving, stable, or declining over time? Investigate anomalies and outliers. Use metrics to drive conversations about improvement, not to punish teams.

Case Example: A Maturity Journey

To make the maturity journey concrete, consider a case example: a mid-size healthcare organization's three-year progression from ad hoc to defined maturity.

Year 1 (Ad hoc → Emerging):
The organization had 12 AI use cases in production (patient scheduling, billing optimization, clinical documentation) but no formal governance. A data breach incident involving training data prompted executive concern. A new Chief Data Officer was hired and tasked with building AI governance. Actions: formed a cross-functional AI Governance Working Group, drafted core policies (data governance, responsible AI principles, approval process), piloted governance with one high-risk use case (clinical risk prediction), conducted risk assessments for all 12 existing use cases, launched training workshops on responsible AI for data science and engineering teams. Results: All use cases risk-assessed and classified. Policies approved by legal and executive team. Three use cases paused for remediation (insufficient documentation, fairness issues). By end of year, organization transitioned to Stage 2 (Emerging).

Year 2 (Emerging → Defined):
Leadership committed resources: hired a full-time AI Governance Manager, secured budget for tools. Actions: formalized AI Governance Committee (monthly meetings, clear charter), implemented model registry with approval workflows, built CI/CD pipelines with automated fairness and security testing, standardized processes (risk assessment template, approval checklist, testing requirements), expanded training (mandatory for all AI practitioners, role-specific modules), brought all 12 existing use cases into compliance (completed documentation, testing, approvals). Results: All use cases compliant. Six new use cases

launched under governance framework. Zero incidents related to AI. Training completion: 95 percent. By end of year, organization transitioned to Stage 3 (Defined).

Year 3 (Defined → Embedded, in progress): Focus shifted to automation and culture. Actions: implemented policy as code (fairness tests, data validation automated in pipelines), deployed continuous monitoring for all production models (drift detection, performance tracking), conducted quarterly governance retrospectives and incorporated lessons learned, launched responsible AI ambassador program (identified champions in each team), integrated responsible AI into performance reviews and promotion criteria. Results: Governance overhead reduced by 40 percent due to automation. Developer satisfaction with governance improved (survey: 4.1/5, up from 2.8/5 in Year 1). One incident detected and resolved within hours (drift in scheduling model, automatically rolled back). Organization is mid-transition to Stage 4 (Embedded), expected to reach it within 12–18 months.

Key success factors: Executive sponsorship and sustained commitment, dedicated resources (people and budget), incremental progress (pilot, standardize, automate), and focus on enablement, not enforcement (training, tools, support).

11.10 Playbook: Leading Cultural Transformation

You have learned the maturity stages, cultural foundations, and progression strategies. Now, how do you lead this transformation? Here is a practical playbook for IT managers.

Step 1: Assess and communicate current state. Conduct maturity assessment (Section 10.3). Share results transparently with leadership and teams. Be honest about gaps. Frame assessment as baseline, not criticism.

Step 2: Define vision and goals. Where do you want to be in 1 year? 3 years? Be specific: "Reach Stage 3 (Defined) within 18 months" or "Automate 80 percent of governance checks within 2 years." Align goals with organizational strategy. Communicate vision clearly and repeatedly.

Step 3: Secure executive sponsorship. Make the case for governance (risks, regulatory pressure, trust imperative). Request resources (budget, staff, executive sponsor). Show early wins. Keep executives informed and engaged.

Step 4: Build coalition of champions. Identify willing teams and individuals who will pilot guardrails and advocate for them. Empower champions with authority, resources, and visibility. Celebrate their successes publicly. Use champions to influence peers.

Step 5: Execute stage-appropriate actions. Use roadmaps from Sections 10.5–10.7. Focus on foundational elements first (policies, structure, processes), then scale and automate. Be patient—maturity takes years, not months.

Step 6: Invest in culture. Do not rely solely on policies and tools. Build psychological safety. Align incentives. Train continuously. Recognize responsible behaviors. Model the culture you want. Culture is built through consistent actions over time.

Step 7: Measure and communicate progress. Track leading and lagging indicators. Report regularly. Celebrate milestones.

Share learnings from incidents and near-misses. Make progress visible to maintain momentum.

Step 8: Adapt and iterate. Governance is not static. Learn from experience. Adjust processes based on feedback. Stay current on regulatory and technology trends. Continuously improve.

Step 9: Sustain over leadership transitions. Institutionalize governance so it survives personnel changes. Document everything. Rotate roles. Tie governance to organizational strategy and structure.

Leading Cultural Transformation Playbook

Lead with vision, reinforce behaviors, and measure progress to shift culture.

11.11 AI Reality Check

Before we close, let's address common myths about maturity and culture:

Myth 1: "Maturity is about having all the right policies and tools."

Reality: Maturity is about culture and behavior. You can have perfect policies and tools but still be at Stage 1 if they are not followed. Culture determines maturity.

Myth 2: "We can skip stages and jump directly to embedded."
Reality: Stages are sequential. You cannot embed what is not defined. You cannot define what does not yet exist. Trying to skip stages leads to superficial compliance without substance.

Myth 3: "Once we reach Stage 4, we are done."
Reality: Stage 4 is not an end state—it is an ongoing practice. Continuous improvement means continuously evolving. Regulations change, threats evolve, and organizations grow. Maturity requires perpetual attention.

Myth 4: "Cultural change is HR's job, not IT's."
Reality: Culture change is everyone's job, especially leaders. As an IT manager leading AI initiatives, you shape culture every day through decisions, priorities, and behaviors. Own it.

Use this reality check to set realistic expectations and maintain focus on what truly drives maturity.

11.12 Summary: Guardrails as AI enabler

This chapter has shown you how to transform AI guardrails from bolted-on controls into cultural DNA. You learned the four stages of AI governance maturity—ad hoc, emerging, defined, and embedded—and how to assess your current state. You learned the five cultural foundations that sustain guardrails: leadership commitment, psychological safety, incentive alignment, education and enablement, and visible recognition. You learned stage-specific strategies for advancing maturity, tactics for preventing backslide, metrics for measuring progress, and a playbook for leading cultural transformation.

The journey from ad hoc to embedded maturity is long—typically 3–5 years of sustained effort. It requires patience, persistence, resources, and unwavering leadership commitment. But the destination is worth it. At Stage 4 (Embedded), guardrails are not obstacles—they are enablers. Teams move faster because risks are managed systematically. Trust is higher because stakeholders see responsible AI in action. Incidents are rare because issues are caught early. Compliance is effortless because it is automated. Responsible AI is not a program—it is identity.

As an IT manager, your legacy is not the systems you build, but the culture you create. When you build guardrails that become invisible, when teams instinctively do the right thing, when responsible AI is how your organization defines excellence—that is when you know you have succeeded. Guardrails are not friction. They are the foundation of trust. And trust is what enables AI to deliver on its promise at scale. Build the guardrails. Build the culture. Make responsible AI who you are.

12 Conclusion

This book has argued a simple but demanding idea: AI guardrails are not optional safeguards for a few exotic systems; they are the operating system for any organization that wants to use AI at scale and stay trusted. As an IT manager, you sit in the uncomfortable but powerful space between ambition and accountability. You are close enough to the technology to know what it can do, and close enough to the business to feel the risk when things go wrong. The chapters you have just worked through are a playbook for turning that tension into a durable advantage.

We started with intent: why guardrails matter now, and how unmanaged AI quietly creates technical, legal, and reputational debt. We then moved into structure—governance models, roles, and decision rights that give you a clear map of who owns what. From there, we translated principles into repeatable practice: policies that actually show up in workflows, technical guardrails that catch problems before they reach users, human-in-the-loop patterns that keep judgment where it belongs, and risk triage models that focus your scarce capacity where it matters most. Monitoring, drift, and incident response brought an operational lens: not "if something goes wrong" but "when it does, how quickly do we detect, contain, and learn."

The second half of the book focused on building an AI program that can stand up under regulatory, organizational, and engineering pressure. You saw how to design compliance once and localize it across jurisdictions, instead of chasing every new rule with a bespoke fix. You embedded guardrails in CI/CD pipelines, model registries, and data workflows so that compliance is enforced by infrastructure, not heroics. Finally, you stepped back to look at culture and maturity: how organizations move from ad hoc controls to an embedded, quietly confident posture where responsible AI is just "how we work here."

As you close this book, the most important next step is not to "do everything" but to pick one or two concrete moves and execute them well. That might be creating a simple AI use-case register and risk triage, standing up a lightweight governance working group, or adding fairness and data-quality checks to a single pipeline. Once that works, expand. Guardrails, like any infrastructure, are built layer by layer: policy, process, tooling, culture. If you keep tying each investment back to mission, value, and risk, you'll find it easier to secure sponsorship and sustain momentum over the long term.

AI will only become more capable, more connected to your core operations, and more visible to regulators and the public. Organizations that treat guardrails as friction will spend the next decade reacting—patching incidents, scrambling for audits, and explaining failures. Organizations that treat guardrails as a strategic asset will move faster with more confidence, because they know their systems are monitored, explainable, and defensible. Your role is to help your organization join the second group. You now have the frameworks, language, and patterns to do that. The rest is leadership: taking the first step, making the hard calls when trade-offs appear, and building a culture where responsible AI is not a slogan, but a habit.